The physics of atmospheres

The physics of atmospheres

JOHN T. HOUGHTON F.R.S.
PROFESSOR OF ATMOSPHERIC PHYSICS
UNIVERSITY OF OXFORD

CAMBRIDGE UNIVERSITY PRESS

CAMBRIDGE
LONDON · NEW YORK · MELBOURNE

6151 - 5061

Published by the Syndics of the Cambridge University Press
The Pitt Building, Trumpington Street, Cambridge CB2 1RP
Bentley House, 200 Euston Road, London NW1 2DB
32 East 57th Street, New York, NY 10022, USA
296 Beaconsfield Parade, Middle Park, Melbourne 3206, Australia

© Cambridge University Press 1977

First published 1977

Printed in Great Britain
at the University Printing House, Cambridge

Library of Congress Cataloguing in Publication Data
Houghton, John Theodore.
The physics of atmospheres.

Bibliography: p.
1. Atmosphere. I. Title.
QC880.H68   551.5′01′53   76-26373
ISBN 0 521 21443 2

To Janet and Peter

'The works of the Lord are great,
sought out of all them that have
pleasure therein.'

Psalm 111, v. 2.

Contents

Preface		*page*	xi
Acknowledgements			xiii
1	Some basic ideas		1
	1.1 Planetary atmospheres		1
	1.2 Equilibrium temperatures		2
	1.3 Hydrostatic equation		3
	1.4 Adiabatic lapse rate		3
	1.5 Sandström's theorem		4
	Problems		6
2	A radiative equilibrium model		8
	2.1 Black-body radiation		8
	2.2 Absorption and emission		8
	2.3 Radiative equilibrium in a grey atmosphere		10
	2.4 Radiative time constants		13
	2.5 The greenhouse effect		14
	Problems		15
3	Thermodynamics		17
	3.1 Entropy of dry air		17
	3.2 Vertical motion of saturated air		17
	3.3 The tephigram		20
	3.4 Total potential energy of an air column		20
	3.5 Available potential energy		22
	3.6 Zonal and eddy energy		26
	Problems		27
4	More complex radiation transfer		31
	4.1 The integral equation of transfer		31
	4.2 Integration over frequency		32
	4.3 Heating rate due to radiative processes		33
	4.4 Single lines		33
	4.5 Transmission of an atmospheric path		37

Contents

4.6 Cooling by carbon dioxide emission from upper stratosphere and lower mesosphere   37

4.7 Absorption of solar radiation by ozone   38

4.8 Band models   39

4.9 Continuum absorption   40

4.10 Global radiation budget   40

     Problems   42

5 The upper atmosphere   46

5.1 Upper atmospheric temperature structure   46

5.2 Diffusive separation   47

5.3 The escape of hydrogen   50

5.4 The energy balance of the thermosphere   53

5.5 Photochemical processes   55

5.6 Breakdown of thermodynamic equilibrium   57

     Problems   63

6 Clouds   67

6.1 Cloud formation   67

6.2 The growth of cloud particles   67

6.3 The radiative properties of clouds   69

6.4 Radiative transfer in clouds   70

     Problems   72

7 Dynamics   74

7.1 Total and partial derivatives   74

7.2 Equations of motion   74

7.3 The geostrophic approximation   77

7.4 Cyclostrophic motion   79

7.5 Surface of constant pressure   79

7.6 The thermal wind equation   80

7.7 The equation of continuity   81

     Problems   82

8 Atmospheric waves   87

8.1 Introduction   87

8.2 Sound waves   87

8.3 Gravity waves   88

8.4 Rossby waves   93

8.5 The vorticity equation   95

8.6 Three dimensional Rossby-type waves   96

     Problems   98

Contents

9	Turbulence	103
	9.1 The Reynolds number	103
	9.2 The Reynolds stresses	104
	9.3 Eckman's solution	105
	9.4 The mixing-length hypothesis	107
	9.5 Eckman pumping	108
	9.6 The spectrum of atmospheric turbulence	109
	Problems	110
10	The general circulation	115
	10.1 Laboratory experiments	115
	10.2 A symmetric circulation	116
	10.3 Inertial instability	121
	10.4 Barotropic instability	122
	10.5 Baroclinic instability	123
	10.6 Sloping convection	127
	10.7 Energy transport	128
	10.8 Transport of angular momentum	129
	Problems	130
11	Numerical modelling	134
	11.1 A barotropic model	134
	11.2 Baroclinic models	134
	11.3 Primitive equation models	136
	11.4 Inclusion of orography	138
	11.5 Convection	138
	11.6 Moist processes	138
	11.7 Radiation transfer	139
	11.8 Sub grid scale processes	143
	11.9 Transfer across the surface	143
	11.10 Other models	144
	Problems	145
12	Global observation	146
	12.1 What observations are required?	146
	12.2 Conventional observations	146
	12.3 Remote sounding from satellites	147
	12.4 Remote sounding of atmospheric temperature	148
	12.5 Remote sounding of composition	152
	12.6 Observations from remote platforms	155
	Problems	157

Contents

13 Atmospheric predictability and climatic change 160
 13.1 Short-term predictability 160
 13.2 Longer-term variations 161
 13.3 Atmospheric feedback processes 162
 13.4 Climate modelling 163

Appendices 164
 1 Some useful physical constants and data on dry air 164
 2 Properties of water vapour 165
 3 Atmospheric composition 166
 4 Relation of geopotential to geometric height 167
 5 Model atmospheres (0–105 km) 167
 6 Mean reference atmosphere 110–500 km 176
 7 The Planck function 176
 8 Solar radiation 177
 9 Absorption of solar radiation by oxygen and ozone 179
 10 Spectral band information 180

Bibliography 187

References to works cited in the text 190

Answers to problems and hints to their solution 193

Index 198

Preface

During the last ten years or so, three important factors have combined to bring about a large increase of interest in atmospheric science. First, man's increasing industrial activity has brought into much sharper focus the problems of pollution and the possibility of artificial modification of the environment either inadvertently or in a controlled way. Secondly, concern about world food resources in the face of rapidly increasing population has made us much more aware of the critical effect of fluctuations in climate, particularly in parts of the developing world. Thirdly, the advent of the artificial satellite and developments in electronic computers have made available much more powerful tools for atmospheric research and have made it possible to study the atmosphere as a whole.

At the present time, study of the global atmosphere is being concentrated by a large international programme – the Global Atmospheric Research Programme – which is being organized jointly by atmospheric scientists (through the International Council of Scientific Unions) and the national weather services (through the World Meteorological Organization). The aims of this programme are to attack some of the basic problems regarding our understanding of the behaviour and circulation of the whole atmosphere (hence the term global) and of the mechanisms of climatic change.

It is in the context of this global programme that this book has been written. Its aim is to introduce physics students at both undergraduate and graduate levels to the physical processes which govern the structure and circulation of a planetary atmosphere. In writing the book I have concentrated on the basic physics and wherever possible have constructed simple physical models by applying to the atmosphere the principles of classical thermodynamics, radiation transfer and fluid mechanics.

The book is meant to be a working text and for that reason includes a large number of problems. Some of the problems extend considerably the basic material in the chapters; other problems are included to illustrate what is already in the text.

I have attempted to introduce all the main areas of study involved in the physics of the neutral atmosphere (the ionized atmosphere and the magnetosphere have been deliberately omitted) while at the same time keeping the book within reasonably small compass. This has necessitated a rigorous selection of material. I hope, however, I have achieved a balanced text from which the reader will be able to acquire a perspective of the atmosphere as a whole and a feeling for where the subject is going. A substantial bibliography is included to help the student to pursue further reading. Also included are a number of appendices summarizing information concerning the atmosphere which research workers will find useful for reference.

Because atmospheric science cuts across a number of traditional disciplines in physics and chemistry, symbols nomenclature and units can be even more of a problem than usual. On the question of symbols I have tended to use those familiar in the various subjects rather than use unfamiliar ones in the interests of uniformity. I have, in general, employed SI units, with one notable exception. For pressure I have kept to the use of millibars, which are universally used in operational meteorology, rather than change to Pascals.

I am indebted to many colleagues and students from whom I have received help in writing the book. Some of the ideas and material for the dynamical chapters arose from an admirable series of lectures on Atmospheric Dynamics given in Oxford by Dr R. S. Harwood, who also made valuable comments and criticisms regarding the text of those chapters. Other criticisms and suggestions on various parts of the text were made by Dr C. D. Walshaw, Dr G. D. Peskett, Professor R. Hide, Professor P. A. Sheppard and by graduate students of Oxford University who have used the text and done the problems. Dr C. D. Rodgers, Mrs M. Corney and Mr R. J. Wells assisted in the collection of some of the material for the Appendices, Mr M. J. Wale has assisted with providing the answers to problems and hints to their solutions. Mr D. Wrigley has read through the proofs and provided the index. I also wish particularly to thank Miss C. M. Wagstaff who typed the manuscript, Miss B. Lindner who drew most of the diagrams, and the staff of the Cambridge University Press for their courtesy and assistance during the preparation of the book.

March 1976   J. T. Houghton

Acknowledgements

Permission has been received to use published diagrams as a basis for figures in the text from the following: Professors J. Charney, H. U. Dütsch, R. Hide, A. S. Monin and E. Palmén; Drs M. Berenger, N. Gerbier, R. Hanel, L. Hembree, P. R. Julian, R. A. McClatchey, S. Manabe, P. J. Mason, L. T. Matveev, C. W. Newton, C. Ridley, D. Rodgers, B. Schlachman, J. E. A. Selby, W. L. Smith, D. Vanous, W. M. Washington, and R. T. Wetherald; Academic Press Inc.; Akademie-Verlag, Berlin; Air Force Geophysics Laboratory; American Meteorological Society; D. Reidel Publishing Co.; Israel Programme for Scientific Translations; McGraw-Hill Book Co.; National Aeronautics and Space Administration, U.S.A.; National Oceanographic and Atmospheric Administration, U.S.A.; Optical Society of America; Royal Meteorological Society; Taylor & Francis Ltd.; and the World Meteorological Organization.

1
Some basic ideas

1.1 Planetary atmospheres

The atmosphere of a planet is the gaseous envelope surrounding it. Large
differences exist between the atmospheres of different planets both in chemical composition and physical structure as will be seen from the information
about the atmospheres of our nearest planetary neighbours contained in
table 1.

Table 1

	Mean surface temperature (K)	Surface pressure (atm)[a]	Accn due to gravity (m s^{-2})	Main constituents	
Venus	750	90	8.84	>90% CO_2	Deep clouds complete cover
Earth	280	1	9.81	N_2 78%[b] O_2 21%	~ 50% cover H_2O clouds
Mars	240	0.007	3.76	>80% CO_2	Some very thin H_2O clouds
Jupiter	134[c]	2[c]	26	H_2, He	NH_3 clouds

[a] 1 atm is mean pressure at earth's surface = 1.013×10^5 Pa(N m^{-2}) =
1013 millibars (mb) (the meteorological unit of pressure which will
be employed throughout the book).
[b] See appendix 3 for detailed composition of earth's atmosphere.
[c] At cloud top.

For a complete understanding of an atmosphere we need to know about
its evolution and the processes which have determined its mass and its composition. We also need to know about its physical structure and the distribution
of density, composition and motion within the atmosphere.

In the case of the earth's atmosphere very detailed information is required
in order to predict its state for periods of days or weeks ahead. Further, because of our concern about climatic change, it is necessary to understand the
factors which determine the average state of the atmosphere over periods of
years and centuries.

In this book discussion will be confined to a description of the physical
processes involved in atmospheric study, with application mostly to the

1

earth's atmosphere. Using basic physical principles in thermodynamics, radiation transfer and fluid dynamics, we shall first develop simple models which may be used to give some understanding of the main features of atmospheric structure and behaviour. In later chapters, methods of observing the atmosphere will be described and an indication given of the major problems which are being tackled in atmospheric science at the present time.

1.2 Equilibrium temperatures

A crude estimate of the effective temperature T_e of a planet's surface may be made by equating the solar radiation it absorbs to the infrared radiation it emits as follows.

$$4\pi a^2 \sigma T_e^4 = \pi a^2 (1 - A) F/R^2 \qquad (1.1)$$

where σ is the Stefan–Boltzmann constant, a is the radius of the planet at distance R (astronomical units) from the sun, $F \, (= 1370 \, \text{W m}^{-2})$ is the solar flux on a surface near the earth (i.e. when $R = 1$) normal to the solar beam and A is the *albedo*, i.e. the ratio of reflected to incident solar energy for the whole planet. The right hand side of (1.1) is the absorbed solar radiation and the left hand side the radiation emitted by the planet assuming it behaves as a black body at temperature T_e. Table 2 lists values of the various parameters in (1.1) for some of the planets and compares approximate measured temperatures T_m with those calculated from (1.1)

Table 2

	R	A	T_e(K)	T_m(K)	M_r	Period of rotation (days)
Venus	0.72	0.77	227	230	44	243
Earth	1.00	0.30	256	250	28.8	1.00
Mars	1.52	0.15	216	220	44	1.03
Jupiter	5.20	0.58	98	130	2	0.41

The agreement is good except in the case of Jupiter, for which planet absorption of solar radiation accounts for only about half the energy input required to maintain its observed temperature; the other half must, therefore, be internally generated (cf. problem 1.2).

The effective temperature for Venus listed in table 2 is very different from the temperature at the planet's surface listed in table 1. The very dense and complete cloud cover of Venus is substantially opaque to radiation of all wavelengths shorter than about 1 mm. These clouds act as a radiation blanket preventing almost all radiation emitted from the lower atmosphere from escaping, while allowing a small amount of solar radiation through (see §1.5).

Some basic ideas

This process whereby a high surface temperature can be maintained is sometimes known as the *greenhouse effect*. To a more limited extent than on Venus it is also effective for the earth where the average surface temperature approaches 290 K.

1.3 Hydrostatic equation

Because a planet's atmosphere is in the planet's gravitational field, its density will fall with altitude. Since vertical motion is generally very small, the assumption of static equilibrium is a good starting point. If ρ is the density and p the pressure at altitude z measured vertically upwards from the surface we have

$$dp = -g\rho dz \tag{1.2}$$

Since a planet's atmosphere is generally of total depth small compared with the planet's radius, the acceleration due to gravity g is approximately constant within the atmospheric region (cf problem 1.5).

From the equation of state for a perfect gas of molecular weight M_r (see table 2) and temperature T

$$\rho = \frac{M_r p}{RT} \tag{1.3}$$

where R is the gas constant per mole.

Equation (1.2), therefore, becomes

$$dp/p = -dz/H$$

which on integration gives, for the pressure p at altitude z,

$$p = p_0 \exp\left\{-\int_0^z dz/H\right\} \tag{1.4}$$

where p_0 is the pressure at $z = 0$, and $H = RT/M_r g$, known as the scale height, is the increase in altitude necessary to reduce the pressure by a factor e. For the lower atmosphere of the earth H varies between 6 km at $T = 210$ K to 8.5 km at $T = 290$ K. In the very high atmosphere, the molecular weight is no longer constant and the temperature is much higher, a situation which is discussed in more detail in §5.2.

1.4 Adiabatic lapse rate

The simplest model of a planetary atmosphere we can make is to assume that it is transparent to all radiation, that it contains no liquid particles and that the temperature of its lower boundary is that of the planet's surface whose mean temperature is determined by the simple calculation of §1.2. Consider the vertical motion of a 'parcel' at pressure p, temperature T and of specific

3

volume V within such an atmosphere. We shall assume the atmosphere to be in hydrostatic equilibrium described by (1.2). Gravitational forces and buoyancy forces are, therefore, balanced and neither need be included explicitly in the expression for the first law of thermo-dynamics applied to unit mass which is therefore

$$dq = c_v \, dT + p \, dV \qquad (1.5)$$

where c_v is the specific heat at constant volume. Providing no heat enters or leaves the parcel, the motion is adiabatic and the quantity of heat dq is zero.

Differentiation of the equation of state (1.3) (remembering that $1/V = \rho$) gives

$$p \, dV + V \, dp = R \, dT/M_r$$

$$= (c_p - c_v) \, dT \qquad (1.6)$$

since for a perfect gas $c_p - c_v = R/M_r$ where c_p is the specific heat at constant pressure. On substituting for pdV from (1.6) in (1.5) we have

$$c_p \, dT - V \, dp = dq = 0 \qquad (1.7)$$

which gives, on substituting from (1.2)

$$\frac{dT}{dz} = -\frac{g}{c_p} = -\Gamma_d \qquad (1.8)$$

Γ_d is known as the *adiabatic lapse rate* for a dry atmosphere. For the earth's atmosphere, $c_p = 1005 \text{ J kg}^{-1}$ and $\Gamma_d \simeq 10 \text{ K km}^{-1}$.

If the atmosphere, therefore, is heated by contact with the surface and vertical motion thereby ensues, we may expect a uniform temperature gradient with altitude of 10 K km^{-1}.

It is informative to consider the stability with respect to vertical motion which occurs in an atmosphere with any temperature distribution b in fig. 1.1. If a parcel initially at X rises adiabatically it will follow the dry adiabatic a to X' where it will be surrounded by atmosphere under the conditions of X'', i.e. it will be warmer than the surroundings and will continue to rise. Point X on curve b is, therefore, unstable. By a similar argument point Y is stable. A given temperature gradient dT/dz represents stable or unstable conditions according as $-dT/dz$ is $<$ or $> \Gamma_d$.

1.5 Sandström's theorem

Instead of assuming that the atmosphere is completely transparent to radiation, let us consider an atmosphere which absorbs strongly, the incident solar energy being deposited in the upper atmosphere as is likely, for instance, to be case in the Venus atmosphere where a very deep continuous cloud layer is present. What temperature structure is to be expected in the atmosphere below the absorbing layer? Because increasing temperature with

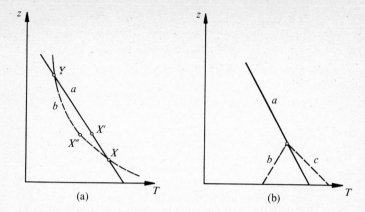

Fig. 1.1. (a) Illustrating the stability of a temperature profile *b*.
(b) Illustrating adiabatic lapse rate *a*; *inversion* condition *b*, i.e. temperature increasing with height, an inversion occurs, for instance, when surface has cooled more rapidly than the atmosphere; *superadiabatic* condition *c* which cannot persist for very long and never occurs on a large scale.

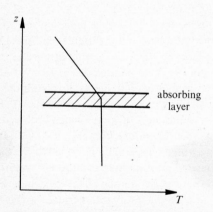

Fig. 1.2. Illustrating an adiabatic lapse rate above and an isothermal state below an absorbing layer.

altitude is a very stable situation (§1.4) vertical motion will not occur; conduction through the gas and radiation transfer will bring the lower atmosphere to an isothermal state (fig. 1.2).

This situation is an illustration of Sandström's theorem (see A. Defant, 1961) which considers motion in the atmosphere to arise from the operation of a thermodynamic engine with heat sources and sinks. In a simple thermodynamic cycle (fig. 1.3), energy is released by going round the cycle in the direction of the arrows. This is only possible if the source is not only at a

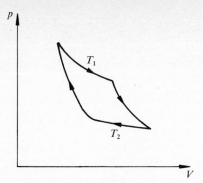

Fig. 1.3. A thermodynamic cycle releases energy when the source at temperature T_1 is at a higher pressure than the sink at temperature T_2.

higher temperature than the sink but also, because for a gas at constant volume pressure increases with temperature, if the source is at a higher pressure than the sink. Hence Sandström's theorem which states that a closed steady circulation can only be maintained in an atmosphere if the heat source is situated at a higher pressure than the heat sink.

Because, in practice, heat is conducted from the precise place where it is deposited to the surrounding atmosphere and the sources and sinks to some extent are distributed by the motions themselves, strict interpretation of the theorem is difficult as Jeffreys (1925) has pointed out. It is, however, certainly true that strong circulations cannot develop unless the conditions of Sandström's theorem are satisfied.

What then about the lower atmosphere of Venus which is observed to possess an adiabatic lapse rate of temperature from the cloud level to the surface which is at the surprisingly high temperature of ~ 750 K? We can only deduce that sufficient solar energy must penetrate to the surface for enough vertical motion to be developed to overcome the effects of conduction and radiation which would tend to make the profile isothermal – a suggestion which is, in fact, in agreement with measurements made by the Russian Venera probes of a small amount of sunlight being still incident on the probes when they reached the Venus surface.

Problems

(See the appendices for lists of constants and useful information about the atmosphere.)

1.1   If the planets of table 2 possessed no atmospheres, what would be the equilibrium temperatures of their surfaces, assuming a uniform albedo of the surfaces of 0.05?

1.2   From the information of table 2 compute the amount of energy per unit area arising from the internal source on Jupiter.

1.3   If the temperature falls uniformly with height z at 10 K per km, write down an expression for the fall of pressure with height.

1.4   Calculate for the earth's atmosphere the height of the surface where the pressure is 0.1 of its surface value, assuming (1) uniform temperature of 290 K (2) surface temperature of 290 K and uniform lapse rate of 10 K km^{-1}.

1.5   What is the percentage change of the acceleration due to gravity g between the surface and 100 km altitude? Estimate the error in pressure determination at 100 km altitude through the use of (1.4) if a constant g is assumed.

1.6   The total mass of the oceans is 1.35×10^{21} kg. Compare this with the total mass of the atmosphere. Also, compare the total heat capacity of the atmosphere with that of the oceans.

1.7   From the information in tables 1 & 2 compute the adiabatic lapse rates for Mars, Venus and Jupiter atmospheres.

1.8   For an unsaturated atmosphere containing 2% by volume of water vapour, find the value of the adiabatic lapse rate and compare it with the value for dry air.

1.9   A balloon filled with helium is required to carry a payload of 100 kg to a height of 30 km. What volume of balloon is required if the material used in its construction is polyethylene sheet, 25×10^{-4} cm thick and of density 1 g cm^{-3}?

1.10   Consider the vertical forces acting on a parcel displaced by a distance Δz from its equilibrium position in an atmosphere where the vertical temperature gradient is dT/dz. Write down the equation of motion in the form

$$\ddot{\Delta z} + gB\Delta z = 0$$

where $B = \dfrac{1}{T}\left(\dfrac{dT}{dz} + \Gamma_d\right)$

Discuss solutions to the equation of motion when $B < 0, = 0, > 0$. When $B > 0$, show that oscillations occur at the *Brunt–Vaisala frequency* $(gB)^{1/2}$. Calculate its value when $dT/dz = -6.5$ km^{-1} – an average value for the lower part of the earth's atmosphere. This oscillation is a particular case of a *gravity wave*. These waves will be discussed further in §8.3.

2
A radiative equilibrium model

2.1 Black-body radiation

In fig. 2.1 are two curves showing the distribution with wavelength of the energy emitted by black-body sources at temperatures of 5750 K, approximating to that of the sun and 245 K, a typical temperature for a planetary surface or a planetary atmosphere. Notice that the two curves are almost entirely separate. Also shown in fig. 2.1 is the absorption by various constituents in the earth's atmosphere; electronic bands occur in the ultraviolet and vibration–rotation and pure rotation bands due to minor constituents in the infrared. The major constituents nitrogen and oxygen, because of their symmetry, possess no electric dipole transitions and hence no strong bands due to them occur in the infrared.

In §1.4 an atmosphere was considered completely transparent to both solar and thermal radiation, in which heat was transferred from the heated surface by vertical convection. Inspection of fig. 2.1 suggests a slightly more elaborate model of an atmosphere transparent to solar radiation so that the surface is heated as before, but possessing an absorption coefficient uniform throughout the emitting infrared region and independent of pressure or temperature. Such an atmosphere is said to be *grey*.

The absorbing constituents in an atmosphere are molecules with complicated absorption spectra, much more complicated in fact than is indicated in fig. 2.1 (b) (cf. fig. 4.2). The approximation of a single absorption coefficient independent of wavelength, pressure and temperature, is a very crude one. However, because of the large amount of overlap of the spectral lines involved, particularly in the lower atmosphere, the approximation is not, in fact, so crude as at first appears and it will enable us to derive features of the basic structure of an atmosphere in which radiative transfer is important. For a more accurate treatment and especially for conditions in the upper atmosphere, more detailed consideration of the spectral structure is necessary. This will be given in chapter 4.

2.2 Absorption and emission

First it is necessary to derive some basic equations in radiative transfer. The equations will be derived for absorption and emission processes only. In the absence of clouds, scattering is unimportant at the infrared wavelengths we

8

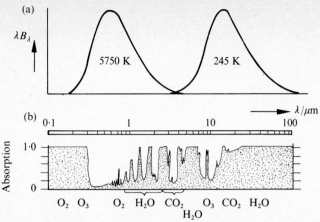

Fig. 2.1.(a) Curves of black-body energy B_λ at wavelength λ for 5750 K (approximating to the sun's temperature) and 245 K (approximating to the atmosphere's mean temperature). The curves have been drawn of equal areas since integrated over the earth's surface and all angles the solar and terrestrial fluxes are equal.
(b) Absorption by atmospheric gases for a clear vertical column of atmosphere. The positions of the absorption bands of the main constituents are marked.

are considering and is, therefore, ignored for the moment. Radiative transfer in clouds is treated in chapter 6; in §6.4 it is shown that scattering processes may easily be incorporated into the same basic equations.

In a plane parallel atmosphere uniform in the horizontal consider first absorption of radiation along the vertical co-ordinate z. The law of absorption (sometimes known as Lambert's law or Bouguet's law) states that the absorption which occurs when radiation of intensity I (units: power per unit area per unit solid angle) traverses an elementary slab of atmosphere of thickness dz is proportional to the mass of absorber ρdz in unit cross-section of the slab (ρ being the density of absorber) and to the incident intensity of radiation itself. Thus

$$dI = -Ik\rho\,dz \qquad (2.1)$$

where k is the absorption coefficient. Integrating (2.1) leads to

$$I = I_0 \exp\left(-\int k\rho\,dz\right) \qquad (2.2)$$

The quantity $\exp\left(-\int k\rho\,dz\right)$ is the *fractional transmission* τ of the path, and $\int k\rho\,dz$ is known as the *optical path* χ (when measured from the top of the atmosphere downwards it is the *optical depth*).

A slab of atmosphere will also emit radiation in amount depending on its temperature. If thermodynamic equilibrium applies, application of

Kirchhoff's law enables the amount emitted per unit area to be written as $k\rho\,dz\,B(T)$ where $B(T)$ is the black-body emission per unit solid angle per unit area of a surface at temperature T.

From the Stefan–Boltzmann law, the integral of $B(T)$ over a hemisphere is proportional to T^4, i.e.

$$\int_{2\pi} B\,dS\cos\theta\,d\omega \;=\; \sigma T^4\,dS$$

giving $\quad B \;=\; \pi^{-1}\sigma T^4$

where σ is the Stefan–Boltzmann constant and $d\omega$ is an element of solid angle at an angle θ to the normal to the element of surface area dS.

The earth's atmosphere is not, of course, precisely in thermodynamic equilibrium; however, in the lower atmosphere conditions known as *local thermodynamic equilibrium* (LTE) prevail which constitute a sufficiently good approximation to equilibrium for black-body emission to be employed in the equation for emission. We shall come back to this point later when considering the high atmosphere where LTE no longer applies (§5.6).

The equation for radiative transfer through the slab, which includes both absorption and emission, is sometimes known as *Schwarzchild's equation*

$$dI \;=\; -I k\rho\,dz + B k\rho\,dz$$

or $\qquad \dfrac{dI}{d\chi} = I - B$ $\hfill (2.3)$

In deriving (2.3) vertically travelling radiation only has been considered. Radiation is, of course, travelling in all directions. However, under the assumption of a plane parallel atmosphere the problem is reduced to a one dimensional one by considering the two fluxes $F^{\uparrow}$ and $F^{\downarrow}$ which are the quantities $\int I(\theta)\cos\theta\,d\omega$ integrated over the downward facing and upward facing hemispheres respectively, $I(\theta)$ being the intensity at an angle θ to the vertical and $d\omega$ an element of solid angle (fig. 2.2). Detailed calculation shows that to a good approximation in (2.3) I may be replaced by F if dz is replaced by a mean thickness $5/3\,dz$ and B is replaced by πB, the black-body function integrated over a hemisphere.

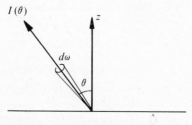

Fig. 2.2

2.3 Radiative equilibrium in a grey atmosphere

Having now laid the theoretical foundation of radiative transfer in a grey atmosphere, we apply it to find the equilibrium situation in an atmosphere in which transfer of infrared radiation is the only energy transfer mechanism and which has for its lower boundary a heated surface at temperature T_g.

Under these assumptions the net rate of temperature change dT/dt which occurs in a slab of plane parallel atmosphere due to the divergence of the upward and downward radiation fluxes is given by (fig. 2.3):

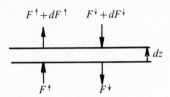

Fig. 2.3.

$$\frac{d}{dz}(F^{\downarrow} - F^{\uparrow}) = \rho c_p \frac{dT}{dt} \tag{2.4}$$

In equilibrium $dT/dt = 0$, and integration of (2.4) gives

$$F^{\uparrow} - F^{\downarrow} = \text{a constant } \phi, \text{ the net flux.} \tag{2.5}$$

Further, from (2.3) the transfer equations are

$$\left.\begin{aligned}\frac{dF^{\uparrow}}{d\chi^*} &= F^{\uparrow} - \pi B \\[2mm] -\frac{dF^{\downarrow}}{d\chi^*} &= F^{\downarrow} - \pi B\end{aligned}\right\} \tag{2.6}$$

where the star on the optical depth χ^*, which is measured from the top of the atmosphere, denotes that it is appropriate to hemispheric radiation (i.e. the factor 5/3 has been applied to dz in the expression for χ).

If

$$\psi = F^{\uparrow} + F^{\downarrow} \tag{2.7}$$

equations (2.6) may be written

$$\frac{d\psi}{d\chi^*} = \phi \tag{2.8}$$

$$\frac{d\phi}{d\chi^*} = \psi - 2\pi B \tag{2.9}$$

Since $\phi = $ constant (from (2.5)) $d\phi/d\chi^* = 0$

and

$$\psi = 2\pi B \tag{2.10}$$

11

which on substitution into (2.8) gives

$$B = \frac{\phi}{2\pi}\chi^* + \text{constant} \tag{2.11}$$

The boundary condition at the top of the atmosphere ($\chi^* = 0$) is $F^\downarrow = 0$, so that here $\psi = \phi$ and, from (2.10), the constant in (2.11) is $\phi/2\pi$, i.e.

$$B = \frac{\phi}{2\pi}(\chi^* + 1) \tag{2.12}$$

At the bottom of the atmosphere where $\chi^* = \chi_0^*$, $F^\uparrow = \pi B_g$, B_g being the black-body function at the temperature of the ground. It is easy to show that there must be a temperature discontinuity at the lower boundary, the black-body function for the air close to the ground being B_0, and

$$B_g - B_0 = \frac{\phi}{2\pi} \tag{2.13}$$

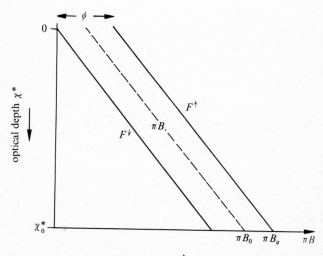

Fig. 2.4. Upward radiation flux $F^\uparrow$, downward flux $F^\downarrow$ and black-body function πB at atmospheric temperature plotted against optical depth for radiative equilibrium atmosphere.

The black-body function B for such an atmosphere is plotted against χ^* in fig. 2.4. The equivalent diagram in terms of temperature and height is fig. 2.5. Notice that the lower part of the atmosphere possesses a very steep lapse rate of temperature, and at the surface itself there is a discontinuity in temperature. As was shown in §1.4, such a steep lapse rate is very unstable with respect to vertical motion, and will soon be destroyed by the process of *convection* which will tend to establish a mean adiabatic lapse rate.

In fig. 2.5 air from near the surface will tend to rise along the line which

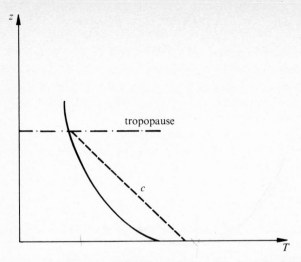

Fig. 2.5. Radiative equilibrium temperature T plotted against altitude z. The line c is drawn through the surface temperature with slope $-6\ \mathrm{K\ km^{-1}}$. This simple model leads to a troposphere dominated by convection below a stratosphere in approximate radiative equilibrium.

has been drawn with a slope of $-6\ \mathrm{K\ km^{-1}}$ (a mean adiabatic lapse rate, see §1.4 and also §3.2) and which cuts the radiative equilibrium curve at the height of approximately 10 km.

   This simple model does, in fact, bear some relation to what occurs in the real atmosphere. The *tropopause* is a surface situated at a height of ~ 10 km in mid latitudes which divides the region below (the *troposphere* or turning-sphere) in which convection is the dominant mechanism of vertical heat transfer, from the *stratosphere* which is much more stably stratified and where radiative transfer is dominant.

2.4 Radiative time constants

In considering radiative processes it is important to know the rate at which changes may be expected to occur. The time constant appropriate to any particular process will depend on the scale of the phenomenon under consideration and the density of the air at the altitude in question as well as on the details of the process itself. For the atmosphere as a whole a rough estimate may be made by considering a slab of atmosphere of thickness h and uniform density ρ radiating like a black body at a temperature $T \simeq 270\ \mathrm{K}$ different by a small amount ΔT from what it would be if in radiative equilibrium with the surrounding layers both above and below the slab.

   Its rate of change of temperature is given by

$$c_p \rho h \frac{d\Delta T}{dt} = 8\sigma T^3 \Delta T \tag{2.14}$$

The time constant resulting from (2.14) is $c_p \rho h / 8\sigma T^3$. If ρ is the density appropriate to the 500 mb level and h equal to the scale height H, i.e. ~ 8 km, the magnitude of this time constant is about 6 days. Radiative processes in the lower atmosphere, therefore, in general, act rather slowly and for the consideration of short-term atmospheric developments may often be neglected. In the longer term, however, because they are the processes which largely determine the distribution of energy sources and sinks they are of dominant importance.

2.5 The greenhouse effect

Combining (2.12) and (2.13) we find that for the radiative equilibrium atmosphere

$$B_g = \frac{\phi}{2\pi} (\chi_0^* + 2) \tag{2.15}$$

where χ_0^* is the optical depth at the bottom of the atmosphere. If $\chi_0^* = 0$, $B_g = \phi/\pi$ and the surface temperature is in equilibrium with the incoming and the outgoing radiation, which are both equal to ϕ. If χ_0^* is large, the surface temperature represented by the black-body function B_g will be very considerably enhanced, an illustration of the so-called *greenhouse effect* mentioned in §1.2.

In practice, the optical depth of an atmosphere will depend on the concentration of gases such as water vapour which are evaporated from the surface and whose concentration may increase rapidly with surface temperature. A mechanism for positive feedback therefore exists which may be particularly important in the case of the Venus atmosphere (Rasool & DeBergh, 1970; Ingersoll, 1969). It has been called the *runaway greenhouse effect* and is illustrated for three planets in fig. 2.6. Suppose the atmospheres of these planets began to form by outgassing from their interiors at a time when their surface temperatures were essentially determined by equilibrium between absorbed solar radiation and emitted long-wave radiation ((1.1) with $A \simeq 0$ because no clouds would be present) at values given by those on the left hand side of fig. 2.6. As water vapour accumulated in an atmosphere, owing to the greenhouse effect the surface temperature rose, so increasing the evaporation from the surface until either the atmosphere became saturated with water vapour or until all the available water had evaporated. The formation of clouds by condensation of water vapour enhances the effect except for the case of Mars where the atmosphere is so thin that significant cloud formation has not occurred and the blanketing effect of the atmosphere is small. For the earth, the equilibrium situation is with most of the water in liquid form, while for Venus, on these assumptions, the surface temperature would always

14

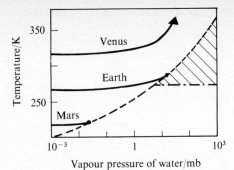

Fig. 2.6. Illustrating the greenhouse effect for the terrestrial planets. Their surface temperature is plotted against the vapour pressure of water vapour in the atmosphere. Also on the diagram (dashed) are the phase lines for water, the shaded area showing where liquid water is in equilibrium. For Mars and the Earth the greenhouse effect is halted when water vapour becomes saturated with respect to ice or water. For Venus the diagram illustrates the 'runaway greenhouse effect'. (After Rasool & De Bergh, 1970 and Goody & Walker, 1972).

be above the boiling point of water at the surface pressure. No liquid water would, therefore, be present on Venus. Supposing that originally similar amounts of water were present as for the earth, the early Venus atmosphere would have water vapour as the major constituent. No other gases would be present to prevent ultraviolet solar radiation dissociating water vapour at the top of the atmosphere; the resulting hydrogen would escape (§5.3) and the oxygen would be consumed in various oxidation processes at the surface. An explanation is, therefore, afforded of the small quantity of water on Venus today compared with the large amount on the earth. The large amount of carbon dioxide remaining in the Venus atmosphere rather than as carbonates in the rocks is also consistent with this atmospheric history.

Problems

2.1   If one third of the solar energy incident outside the atmosphere were absorbed by the atmosphere, show that the average rate of temperature rise (assuming no loss) of the atmosphere would be ~ 1 K per day. Rates of temperature change due to radiative processes are often expressed in units of K per day. For instance an average rate of loss of energy by the troposphere due to emission of long-wave radiation is ~ 1 K per day.

2.2   Assuming an absorber with uniform absorption coefficient and uniformly mixed with height, a surface temperature of 280 K, a tropopause temperature at 250 mb pressure of 220 K and radiative equilibrium conditions, find ϕ. Also find χ^* in equation (2.12) as a function of pressure. What would be the temperature discontinuity at the surface under such assumptions?

2.3   The calculation of §2.3 has been carried out on the assumption that radiative equilibrium exists throughout the atmosphere. Make a qualitative assessment of the effect on the radiative equilibrium profile above the tropopause of a change in the profile below the tropopause from the radiative equilibrium one to the convective one (fig. 2.5).

2.4   Estimate the radiative time constant for the atmosphere of Mars.

2.5   What is the length of a solar day on Venus? (cf. table 2 noting also that the sense of rotation of Venus is opposite to that of the earth and Mars). Ignoring the effect of the clouds, by a similar calculation to problem 2.4, estimate the radiative time constant for the lower Venus atmosphere. Because of the blanketing effect of the clouds and probably also because of motion in the Venus atmosphere, no difference in temperature in the lower Venus atmosphere has been measured between the day and the night side.

2.6   The surface temperature of Venus is $\sim 750\ \mathrm{K}$ (table 1). From the value of T_m from table 2 and from (2.15) assuming radiative equilibrium, calculate the optical depth of the Venus atmosphere.

3
Thermodynamics

3.1 Entropy of dry air

In chapter 1 the first law of thermodynamics was applied to vertical motions in a planetary atmosphere and the adiabatic lapse rate for dry air was derived. On substituting TdS for dq in (1.7) where S is the entropy and substituting for V from the equation of state for a perfect gas we have

$$dS = c_p \frac{dT}{T} - R \frac{dp}{p} \quad \text{per mole !} \tag{3.1}$$

which on integration gives an expression for the entropy

$$S = c_p \ln T - R \ln p + \text{constant} \tag{3.2}$$

If air at pressure p and temperature T is brought adiabatically (i.e. at constant S) to a standard pressure p_0 of 1000 mb, its temperature θ at that standard pressure is known as its *potential temperature*.

From (3.2)

$$c_p \ln \theta = c_p \ln T - R \ln p + R \ln p_0 \tag{3.3}$$

from which, remembering that for a perfect gas $c_p - c_v = R$

$$\theta = T \left(\frac{p_0}{p} \right)^\kappa \tag{3.4}$$

where $(c_p - c_v)/c_p = \kappa = 0.288$ for dry air in the earth's atmosphere.

Also

$$S = c_p \ln \theta + \text{a constant} \tag{3.5}$$

– a useful equation because it enables us to express the entropy of dry air in terms of the more readily interpretable concept of potential temperature.

3.2 Vertical motion of saturated air

The structure of an atmosphere may be influenced very considerably by the presence within it of a condensable vapour. In the case of the earth's atmosphere water vapour is very important.

Consider the vertical motion of air containing water vapour. Providing air remains unsaturated, its thermodynamic properties are little different from

17

those of dry air (problem 3.1). As unsaturated air rises, therefore, it cools similarly to dry air at about $10\,\mathrm{K\,km^{-1}}$. During this ascent although the *mixing ratio** of water vapour remains constant, the *relative humidity*[†] increases and may reach 100%. The level at which this occurs is the *condensation level*. As the air continues to rise, it remains saturated, the surplus water vapour condensing to form drops of liquid water. The latent heat released by this condensation process must now be included.

The same assumptions are made as in §1.4 together with the additional assumption that the parcel contains liquid water which moves with it; the air remains saturated, evaporation and condensation occurring so that equilibrium is continually maintained. Consider a parcel of 1 g of dry air with m g of water vapour (m is the *saturation mixing ratio*) and $\xi - m$ g of liquid water. For the dry air, the first law of thermodynamics is as stated in (1.7). For the water vapour and liquid water, the first law is

$$dq = \xi c\, dT + d(Lm) \tag{3.6}$$

where c is the specific heat of liquid water and L is the latent heat which, of course, varies with temperature. For the complete parcel under adiabatic conditions, therefore, from (1.7) and (3.6)

$$(c_p + \xi c)\, dT + d(Lm) - V\, dp = 0 \tag{3.7}$$

We are dealing with wet air so that the hydrostatic equation becomes

$$dp = -g\rho_a(1 + \xi)\, dz \tag{3.8}$$

From (3.8) and (3.7)

$$(c_p + \xi c)\, dT + d(Lm) + g(1 + \xi)\, dz = 0 \tag{3.9}$$

The mixing ratio m is dependent on air pressure as well as temperature, so it is convenient to use the saturation vapour pressure e as a parameter. The equations of state for unit masses of the air (density ρ_a, molecular weight M_{ra}) and water vapour (density ρ_v, molecular weight M_{rv}) considered separately are

and
$$\left.\begin{array}{r} p - e = \rho_a RT/M_{ra} \\ e = \rho_v RT/M_{rv} \end{array}\right\} \tag{3.10}$$

Now
$$m = \frac{\rho_v}{\rho_a} = \left(\frac{e}{p - e}\right)\frac{M_{rv}}{M_{ra}} \simeq \frac{e\epsilon}{p} \tag{3.11}$$

---

* i.e. the ratio of number of water molecules to other molecules in a given volume (the volume mixing ratio) or the ratio of the mass of water vapour to mass of air in a given volume (the mass mixing ratio).
† the ratio of water vapour pressure to saturation water vapour pressure at air temperature.

since, in the lower atmosphere, $e \ll p$, and where

$$\epsilon = \frac{M_{rv}}{M_{ra}} = 0.622$$

Differentiating (3.11) gives

$$\frac{dm}{dz} = \frac{\epsilon}{p}\frac{de}{dT}\cdot\frac{dT}{dz} - \frac{\epsilon e}{p^2}\cdot\frac{dp}{dz} \qquad (3.12)$$

Substituting from (3.12) in (3.9), considering the air just saturated, i.e. $\xi = m$, and substituting from the Clausius Clapeyron equation

$$\frac{de}{dT} = \frac{LeM_{rv}}{RT^2} \qquad (3.13)$$

we have

$$-\frac{dT}{dz} = \Gamma_s = \frac{g}{c_p}\cdot\frac{\{1+(LeM_{rv}/pRT)\}\{1+(\epsilon e/p)\}}{[1+(\epsilon e/pc_p)\{c+(dL/dT)\}+(\epsilon eL^2M_{rv}/c_p pRT^2)]} \qquad (3.14)$$

On inserting typical values (problem 3.2) the middle term in the denominator on the r.h.s. turns out to be negligible in comparison with the other terms. Also, since $e \ll p$, the second bracket in the numerator $\simeq 1$, so that (3.14) becomes

$$\Gamma_s = \Gamma_d\left(1+\frac{LeM_{rv}}{pRT}\right)\left(1+\frac{LeM_{rv}}{pRT}\cdot\frac{\epsilon L}{c_p T}\right)^{-1} \qquad (3.15)$$

Γ_s is always $< \Gamma_d$ and varies in the atmosphere from about $0.3\Gamma_d$ to Γ_d. Actual values can be read from fig. 3.1.

The stability of saturated air containing liquid water for vertical motion can be considered in exactly the same way as that for dry air in §1.4. We have stable or unstable conditions according as

$$-dT/dz \text{ is } < \text{ or } > \Gamma_s$$

It should be noted that for saturated air with no liquid water content, the stability condition for upward motion involves Γ_s, while that for downward motion involves Γ_d, since as soon as the air moves downward it becomes unsaturated.

The assumption made above that the liquid water remains with the parcel as it is lifted, is clearly not necessarily fulfilled in practice. If, however, it is assumed that the products of condensation, or a proportion of them fall out during the ascent, little difference is made to the final answer. Another assumption which has been made in considering vertical motion is that the environment is unmodified by the rising or sinking parcels of air. The motion of a parcel of air is, of course, bound to produce a compensating motion in the environment. However, since the compensating descending motions normally occur over a much larger area than the ascending motion, the *parcel* method we have described which ignores the compensating motions is in general a good approximation and is used in general forecasting analysis.

3.3 The tephigram

The tephigram is a thermodynamic diagram particularly suitable for representing atmospheric processes. The name derives from T–ϕ-gram since it is a temperature–entropy plot, the symbol ϕ sometimes being used for entropy (we have used the symbol S). Instead of plotting S as the ordinate in our thermodynamic diagram, we can plot $\ln \theta$ (3.5) remembering that θ is the potential temperature referred to *dry* air. The basic diagram therefore is a plot of $\ln \theta$ against T. Contours of other quantities may also be plotted on the tephigram to assist in its use, namely

(1) the pressure p, from (3.4), is a function of T and θ, so isobars or lines of constant pressure can be drawn. For many applications it is then convenient to use the diagram as one in which T can be plotted as a function of p.

(2) Adiabatic lines for dry air are lines of constant θ.

(3) Adiabatic lines for saturated air may be plotted from (3.15) where the lapse rate is given as a function of T and p. These *wet adiabatics* show the entropy change of the saturated air which is exactly equal but opposite in sign to that of the liquid water carried with it (or falling out of it).

(4) The saturation mixing ratio m from (3.11) is a function of T and p only. Note that a plot of mixing ratio against pressure for unsaturated air, is identical to a plot of *dew point* (or frost point for temperatures below zero centigrade) against pressure.

Fig. 3.1 is an example of a tephigram on which the above isopleths are drawn. Problems at the end of the chapter illustrate its use in some detail.

3.4 Total potential energy of an air column

The energy dE of an element of a static air column at altitude z consists of two parts, its internal energy dE_I which is $c_v T$ per unit mass and its potential energy dE_P which is gz per unit mass. Since an element of mass is $-dp/g$, (1.2) we have

$$E = E_I + E_P = g^{-1} \int_{p_h}^{p_0} c_v T \, dp + \int_{p_h}^{p_0} z \, dp \qquad (3.16)$$

where the column extends from $z = 0, p = p_0$ to $z = h, p = p_h$. The second term on the r.h.s. may be integrated by parts to give

$$E_P = -p_h h + \int_0^h p \, dz \qquad (3.17)$$

Now since $p = R\rho T/M_r$ (equation of state (1.3))

Thermodynamics

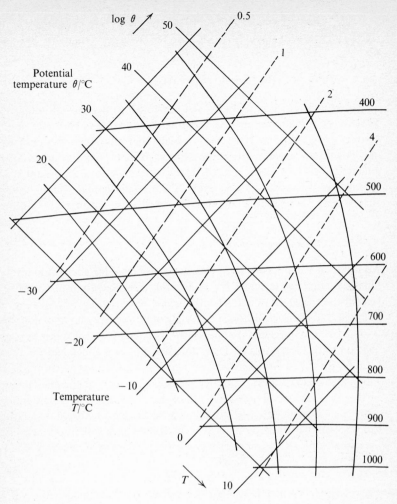

Fig. 3.1. Part of a tephigram. (Charts available in the UK from Her Majesty's Stationery Office.)

The main axes of entropy and temperature are inclined. Isopleths of pressure (marked in mb) are then nearly parallel to the lower edge of the page.

The dashed lines are lines of constant water vapour mixing ratio (labelled in g kg^{-1}) and the most curved lines are wet adiabatics.

and $\rho\,dz = -dp/g$ we have

$$E_P = -p_h h + g^{-1} \int_{p_h}^{p_0} \frac{RT}{M_r}\,dp \qquad (3.18)$$

21

The first term may be made negligible by choosing h sufficiently high that $p_h \simeq 0$. Substituting from (3.18) into (3.16), remembering that $c_p = c_v + R/M_r$ we have

$$E = E_I + E_P = c_p g^{-1} \int_0^{p_0} T \, dp \tag{3.19}$$

Note that $E_P/E_I = c_p/c_v - 1 \simeq 0.4$ for dry air so that, to the extent that hydrostatic equilibrium prevails, the potential and internal energies of a column of air bear a constant ratio to each other. Conversion into kinetic energy will, therefore, occur at the expense of both potential and internal energy, in this same ratio. It is, therefore, convenient to treat potential and internal energy together; their sum is called *total potential energy* – a name first used by Margules in a famous paper in 1903 on the energy of storms.

3.5 Available potential energy

All the total potential energy is clearly not available for conversion into kinetic energy. For instance, for a uniformly stratified atmosphere stable with respect to vertical motion in which there are no variations of density in the horizontal at any level, although the total potential energy is large, none at all is available for conversion. Suppose now that such a stratified atmosphere is heated in a restricted region. Total potential energy is added and the stratification is disturbed; horizontal density gradients and hence pressure forces are created which may convert total potential energy into kinetic energy. Suppose further that a uniformly stratified atmosphere is cooled over a limited region rather than heated. Although total potential energy is removed, as a result of this removal the stratification is disturbed, and conversion into kinetic energy is again possible. Removal of energy can be as effective as addition of energy in making more energy available.

A quantity which for any given atmospheric state is a measure of the total potential energy which is available for conversion through adiabatic processes into kinetic energy was introduced by Lorenz (1955) and is called *available potential energy*. It is the difference between the total potential energy in the state under consideration, and the total potential energy of a state of uniform stratification which is statically stable (i.e. having a stable lapse rate of temperature) obtained by a redistribution of atmospheric mass by adiabatic processes. For conditions of such uniform stratification the available potential energy is zero.

The process of readjustment to the statically stable reference state involves imposing vertical motions so that the originally undulating isentropic surfaces (i.e. surfaces of constant potential temperature) are brought into coincidence with surfaces of constant geopotential (cf. §7.5 and problem 7.11 for definition of geopotential). Since warmer air rises and cooler air descends during the readjustment gravitational potential energy will be released and work will

Thermodynamics

be done at the expense of internal energy; the sum of these is the available potential energy we wish to find.

To calculate its magnitude, following Lorenz (1955), consider the atmosphere divided up by surfaces of constant potential temperature θ (§3.1) such as those illustrated by the example of fig. 3.2.

In any redistribution under adiabatic flow the value of θ for any given element of atmosphere will be conserved. Providing that each of the surfaces of constant θ intersect every vertical column only once, we can easily define the average pressure $\bar{p}$ over each of the surfaces, with respect to the area projected on to a horizontal surface. Surfaces which appear to intersect the ground may be imagined to continue at ground level, i.e. where $p = p_0$. The total potential energy E of a vertical column (3.19) can be written in terms of potential temperature θ from (3.4) and integrated by parts to give

$$E = (1 + \kappa)^{-1} c_p g^{-1} p_0^{-\kappa} \left(\int_{\theta_0}^{\infty} p^{1+\kappa} d\theta + \theta_0 p_0^{1+\kappa} \right) \tag{3.20}$$

Providing that the value of θ at the surface, namely θ_0, is lower than any value in the atmosphere above (this is a requirement for stability; see also fig. 3.2), the last term on the r.h.s. of (3.20) can be incorporated in the integral by replacing the lower limit with $\theta = 0$ with the interpretation that all surfaces of θ with values less than θ_0 coincide at $p = p_0$. We therefore have for the whole atmosphere (per unit area of surface)

$$E = (1 + \kappa)^{-1} c_p g^{-1} p_0^{-\kappa} \int_0^{\infty} \overline{p^{1+\kappa}} \, d\theta \tag{3.21}$$

where $\overline{p^{1+\kappa}}$ is the average of $p^{1+\kappa}$ over an isentropic surface, the average to be taken in the same way as when defining the average pressure $\bar{p}$.

The minimum total potential energy which can result from adiabatic rearrangement occurs when $p = \bar{p}$ everywhere over a given isentropic surface. The difference between this minimum value of total potential energy and the original value is the *available potential energy* E_A and is, therefore, given by

$$E_A = (1 + \kappa)^{-1} c_p g^{-1} p_0^{-\kappa} \int_0^{\infty} (\overline{p^{1+\kappa}} - \bar{p}^{1+\kappa}) \, d\theta \tag{3.22}$$

Because $1 + \kappa > 1$, $\overline{p^{1+\kappa}} > \bar{p}^{1+\kappa}$ unless $p \equiv \bar{p}$, so that (3.22) ensures that E_A is positive, as is, of course, necessary.

In (3.22) let $p = \bar{p} + p'$, $p^{1+\kappa}$ can then be expanded by the binomial theorem. Remembering that $\overline{p'} = 0$ and expanding to the first term only

$$E_A = \tfrac{1}{2} \kappa c_p g^{-1} p_0^{-\kappa} \int_0^{\infty} \bar{p}^{1+\kappa} \overline{\left(\frac{p'^2}{\bar{p}} \right)} d\theta \tag{3.23}$$

Equation (3.23) expresses E_A in terms of the variance of p on an isentropic surface. It is often more convenient to refer to isobaric surfaces, in which case an appropriate expression for E_A is (problem 3.12)

23

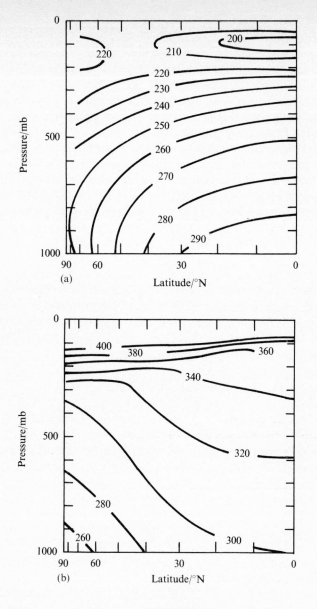

(a)

Latitude/°N

(b)

Latitude/°N

$$E_A = \tfrac{1}{2} c_p g^{-1} \int_0^\infty \bar{T} \left(1 - \frac{\Gamma}{\Gamma_d}\right)^{-1} \overline{\left(\frac{T'}{\bar{T}}\right)^2} \, dp \qquad (3.24)$$

Equation (3.24) is a useful form for estimating E_A. By comparing it, for instance, with (3.19), an estimate can be made of the ratio of available potential energy E_A to total potential energy E. Typical values in (3.24) are

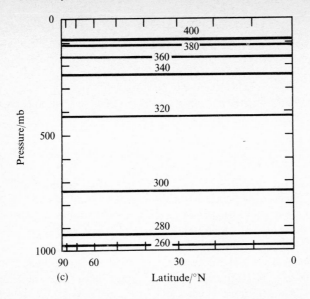

Fig. 3.2. (a) Average temperature cross-section of atmosphere for northern hemisphere winter. (b) Distribution of potential temperature θ corresponding to temperature distribution in (a). (c) Potential temperature distribution which results when atmosphere shown in (b) is moved isentropically to a state of uniform horizontal stratification. (After Lorenz, 1967)

$\Gamma \simeq \frac{2}{3}\Gamma_d$ and $\overline{(T'/\overline{T})^2} \simeq (\frac{1}{16})^2$ giving $E_A/E \simeq 1/200$. Only about $\frac{1}{2}\%$ of the total potential energy, therefore, is available for conversion into kinetic energy.

It is interesting to compare the average kinetic energy with both the total potential energy and the available potential energy. The total potential energy of a vertical column (3.19) may be written as

$$E = \{(\gamma - 1)g\}^{-1} \int_0^{p_0} c^2 \, dp \qquad (3.25)$$

where $\gamma = c_p/c_v$ and $c = (\gamma RT)^{1/2}$ is the speed of sound in air. The kinetic energy E_K of a vertical column expressed in a similar form is

$$E_K = \frac{1}{2g} \int_0^{p_0} V^2 \, dp \qquad (3.26)$$

where V is the velocity of atmospheric motion. A typical value of V is $\sim 0.05c$ so that the ratio E/E_K may be ~ 2000 on average.

Since $E/E_A \simeq 200$, $E_K/E_A \simeq 0.1$ on average, so that the amount of available potential energy in the atmosphere is much larger than the amount of kinetic energy.

The concept of available potential energy may be employed to investigate the efficiency of heating or cooling different parts of the atmosphere in contributing to the maintenance of its general circulation. If the quantity $\dot{q}\,dm$ (dm being an element of mass) represents the rate at which heat is introduced or removed by diabatic processes (e.g. radiation, latent heat, turbulent transfer from the surface etc.), it can be shown that (see for instance Dutton & Johnson (1967)) the rate of generation of available potential energy is approximately

$$\int \left\{ 1 - \left(\frac{\bar{p}}{p}\right)^{\kappa} \right\} \dot{q}\,dm \tag{3.27}$$

Lorenz called the quantity $\{1 - (\bar{p}/p)^{\kappa}\}$ the efficiency factor. Inspection of fig. 3.2 (problem 3.14) shows that maximum generation of available potential energy will occur with low altitude heating in low latitudes and high altitude cooling in polar regions – a result which might be expected from the very general thermodynamic arguments of §1.5.

In the above discussion the available potential energy has been computed by reference to a *statically stable* reference state. Since there are dynamical constraints on the motions which may occur in the atmosphere (see chapters 7ff) it is not clear that processes for attaining such a reference state are dynamically possible. In fact, as has been suggested by Van Mieghem (see Dutton & Johnson, 1967), it may be more realistic to postulate a reference state which includes a circumpolar vortex balanced by a horizontal temperature gradient which may be reached by isentropic flows in the manner described above, but which also preserves the absolute zonal angular momentum.

3.6 Zonal and eddy energy

The above discussion has not been concerned at all with the way in which the conversion from potential energy to kinetic energy takes place. In chapter 10, two mechanisms for such conversion will be described, namely (1) through mean motions which may be considered independent of longitude, and (2) through large scale quasi-horizontal eddy motions which show large variations with longitude. For studies of the general circulation, therefore, it is convenient to divide the kinetic energy into two components by resolving atmospheric motion into (1) the mean zonal component, i.e. the wind velocity averaged around latitude circles, and (2) the eddy components, i.e. the meridional (or north–south) component and the deviation of the zonal component at any place from the mean around the latitude circle. These components are known as the zonal kinetic energy and eddy kinetic energy respectively.

In a similar way zonal and eddy available potential energy may be defined. Components of the variance of the temperature field are chosen, namely (1) the variance of the zonally averaged temperature and (2) the variance of

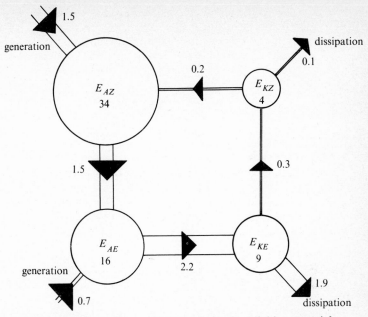

Fig. 3.3. Average storage and conversions of available potential energy E_A and kinetic energy E_K for the whole atmosphere as estimated by Oort & Peixoto (1974). Units of energy in $10^5\,\mathrm{J\,m^{-2}}$ and of conversion, generation or dissipation in $\mathrm{W\,m^{-2}}$. Subscripts Z and E are for zonal and eddy components respectively.

temperature along latitude circles, from which zonal and eddy components of available potential energy may be deduced.

Fig. 3.3 shows an example of the result of a study of energy generation and conversion. Notice that most of the conversion to kinetic energy occurs from the eddy available potential energy which in turn is converted through large scale eddy motions from zonal available potential energy.

Problems

3.1 To allow in a convenient way for the different molecular weight M_{rw} which applies to air containing water vapour, a *virtual temperature* T^* is defined such that

$$RT^*/M_{ra} = RT/M_{rw}$$

where M_{ra} is the molecular weight of dry air. Derive an expression for the virtual temperature in terms of water vapour mixing ratio and find $T^* - T$ for saturated air at 273 K, 290 K and 300 K.

3.2 Calculate typical values of the middle term in the denominator of (3.14) and show that for all atmospheric conditions, it is $\ll 1$. [dL/dT is approximately $-2\,\mathrm{J\,g^{-1}\,K^{-1}}$.]

27

3.3   Calculate values of the term $\epsilon L/c_p T$ in (3.15) and show that because it is always > 1, $\Gamma_s < \Gamma_d$.

3.4   From the hydrostatic equation (1.2) show that the pressure $p(z)$ at height z is related to the field of potential temperature $\theta(z')$ by the equation

$$\left\{\frac{p(z)}{p(0)}\right\}^{\kappa} = \frac{\kappa g M_r}{R} \int_z^{\infty} \frac{dz'}{\theta(z')}$$

where $\kappa = (\gamma - 1)/\gamma$

3.5   The following measurements were made at 0001 GMT on a day in June by a radiosonde ascent from Liverpool:

Pressure (mb)	Temperature (°C)	Dew (or frost) point (°C)
1000	13	11
940	9.5	8
900	7	5
780	0	-3
700	-5	-11
600	-11	-17
500	-20	-28
400	-32	-42
300	-47	
200	-49	
150	-50	
100	-48	

Plot these on a tephigram chart and answer the following questions:
(1)   What is the pressure at the tropopause?
(2)   What parts of the ascent are stable for (a) dry air, (b) saturated air?
(3)   What is the mixing ratio, water vapour:air, at 1000 mb, 500 mb?
(4)   If the surface cools radiatively during the night, how many degrees of cooling are required for fog to begin to form?
(5)   If air at the surface is heated during the following day and then rises adiabatically, at what level will condensation occur?
(6)   What will be the level of the top of convective clouds which develop during the day?

3.6   Air represented by the measurements given in problem 3.5 is forced to ascend either over a range of mountains or over a wedge of colder air at a cold front. Consider the layer of air initially between 780 and 700 mb; note that it is stable for both dry and wet ascent. Suppose this air is forced to rise by 100 mb in pressure. Note that it will remain 80 mb deep in pressure units. Using a tephigram chart find

28

the new temperature of the bottom of the ascent by taking it up the dry adiabatic until it becomes saturated, then up the wet adiabatic. Do the same for the top of the ascent. Hence show that the layer is now unstable with respect to wet ascent. Such a situation is called one of *potential instability*; the amount of such instability will determine how much convective development will occur.

Carry out the same exercise for the layer between 900 mb and 780 mb.

3.7 Again considering the ascent in problem 3.5, take air from the surface up the dry adiabatic to its condensation level, then up the wet adiabatic until it reaches the temperature of the environment. By measuring the appropriate area on the diagram calculate the work required to raise 1 kg of air to this point.

Now continue the ascent along the wet adiabatic until the parcel again reaches the temperature of the environment. Calculate the energy released in this second ascent.

Show that for the complete ascent the net energy release is positive. Such a condition is known as *latent instability*. Once the surface has been heated sufficiently to initiate vertical motion, energy can be released and intense convective activity and maybe thunderstorms are likely to develop.

3.8 Two parcels of damp air at the same pressure with masses M_1 and M_2, water vapour mixing ratios m_1 and m_2, and temperatures T_1 and T_2 are mixed. Show that the resulting temperature and mixing ratio are given by

$$\bar{T} = \frac{M_1 T_1 + M_2 T_2}{M_1 + M_2}$$

$$\bar{m} = \frac{M_1 m_1 + M_2 m_2}{M_1 + M_2}$$

(remember $m \ll 1$),

Using data available on the tephigram chart consider the mixing of two equal masses at 1000 mb pressure and 24°C and 12°C respectively. What relative humidity (assumed the same for each mass) must the masses possess for condensation to occur on mixing the masses together.

3.9 (1) Consider the vertical mixing of a column of air extending from pressure p_1 to pressure p_2. Show that the mixing ratio $\bar{m}$ of the mixed column is given by

$$\bar{m} = \frac{1}{p_2 - p_1} \int_{p_1}^{p_2} m \, dp$$

where $m(p)$ describes the initial distribution.

Show also with appropriate assumptions (state what they are) that an approximate expression for the potential temperature $\bar{\theta}$ of the mixed column is

$$\bar{\theta} = \frac{1}{p_2 - p_1} \int_{p_1}^{p_2} \theta \, dp$$

(2) Plot the following on a tephigram chart: 1000 mb, 28 °C; 950 mb, 24 °C; 900 mb, 21 °C. The water vapour mixing ratios are 20 g kg^{-1} from 1000 mb to 950 mb and 17 g kg^{-1} from 950 mb to 900 mb. If the 1000 mb to 900 mb layer is mixed, estimate the resulting potential temperature and the condensation level. Your result demonstrates that mixing processes can lead to cloud or fog formation.

3.10   Plot on a tephigram chart the following large scale circulation cell between the equator and 30 °N. Air rises adiabatically from 1000 mb and 25 °C to 800 mb when it becomes saturated and continues along the saturated adiabatic to 185 mb and − 75 °C. It then moves northward and sinks losing heat radiatively to 280 mb and − 72 °C after which it descends adiabatically to the surface and returns along the surface while being heated to its starting point. Calculate the energy released in the cycle by unit mass of air.

    If the overturning takes 6 months compare the rate of release of energy with the average input of solar radiation.

3.11   From the information contained in fig. 3.2 carry out a numerical integration of (3.22) and find the available potential energy in the northern hemisphere. Note that since fig. 3.2 contains zonally averaged information your answer refers to zonal available potential energy. Compare your result with that in fig. 3.3.

3.12   Derive (3.24) from (3.23) as follows:

(1)   If on an isobaric surface $\theta = \bar{\theta} + \theta'$, $\bar{\theta}$ being an average value over the surface, show that on a neighbouring isentropic surface where $p = \bar{p} + p'$

$$p' = \theta' \, dp/d\theta$$

(2)   From (3.4) show that

$$\frac{d\theta}{dp} = -\kappa \theta p^{-1} \left(1 - \frac{\Gamma}{\Gamma_d} \right)$$

where $\Gamma = -dT/dz$ is the local lapse rate of temperature and $\Gamma_d = g/c_p$ is the adiabatic lapse rate.

(3)   Make the assumption that in the result of part (1) the average vertical stability may be employed, i.e. $p' = -\theta' \, d\bar{p}/d\theta$, and substitute from parts (1) and (2) in (3.23) to derive (3.24).

3.13   Compare the average input of solar radiation over the hemisphere with the following quantities in fig. 3.3:

(1)   rate of generation of zonal available potential energy,

(2)   the rate of generation of kinetic energy.

3.14   From the information in fig. 3.2 plot a cross-section of the efficiency factor for generation of available potential energy (see (3.27)).

4

More complex radiation transfer

4.1 The integral equation of transfer

In discussing radiation transfer in chapter 2 a very simplified atmospheric model was presented. The assumptions were made of an atmosphere in radiative equilibrium and of an absorption coefficient independent of frequency. In this chapter we shall set up the problem more generally showing how, for any given atmosphere, integrations over height and over frequency can be carried out. We shall then consider a particular case, but still confining the discussion to absorption and emission processes only, i.e. to a non-scattering atmosphere.

To evaluate the effect of radiative transfer in the atmosphere we need to be able to find the intensity of radiation at any given level in an atmosphere with any structure and composition. For the plane parallel atmosphere considered in §2.2 an expression for the intensity is obtained by integrating Schwarzschild's equation (2.3). This may readily be done using the integrating factor $\exp(-\chi)$ (problem 4.1). The integral equation may also be derived directly in a way which illustrates the physical processes involved. Consider a thick slab (fig. 4.1) between levels z_0 and z_1, with radiant intensity I_0 incident vertically upwards at z_0. To calculate I_1, the intensity leaving the top of the slab, consider the elemental slab dz at level z ($z_0 < z < z_1$) and temperature T emitting radiation in the upward direction of intensity $k\rho\,dz\,B(z)$ where the definition of the quantities is as in §2.2. This emitted radiation will be attenuated before reaching z_1, the transmission (§2.2) being

$$\tau(z, z_1) = \exp\left(-\int_z^{z_1} k\rho\,dz'\right) \tag{4.1}$$

The contribution to I_1 from the slab dz will be

$$dI_1 = k\rho\,dz\,B(z)\exp\left(-\int_z^{z_1} k\rho\,dz'\right)$$

$$= B(z)\,d\tau(z, z_1) \tag{4.2}$$

as may easily be verified by differentiating (4.1). We have, therefore

$$I_1 = I_0\,\tau(z_0, z_1) + \int_{\tau(z_0,\,z_1)}^{1} B(z)\,d\tau(z, z_1) \tag{4.3}$$

31

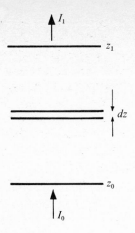

Fig. 4.1

the first term being the contribution from the incident intensity at z_0.

Equation (4.3) is the integral equation of transfer and is important not only in radiation transfer within the atmosphere but also in the interpretation in terms of atmospheric structure of measurements of radiation leaving the top of the atmosphere (§12.4).

4.2 Integration over frequency

Following the treatment of chapter 2 the assumption has so far been made that the absorption coefficient, and hence the transmission τ in (4.3) is independent of frequency – the grey atmosphere approximation. To improve on this approximation, a proper integration over frequency should be carried out to make allowance for the complex line structure of molecular bands so that (4.3) becomes

$$I_1 = \int_0^\infty I_{0\tilde{\nu}}\tau_{\tilde{\nu}}(z_0, z_1)\, d\tilde{\nu} + \int_0^\infty \int_{\tau_{\tilde{\nu}}(z_0, z_1)}^1 B_{\tilde{\nu}}(z)d\tau_{\tilde{\nu}}(z, z_1)d\tilde{\nu} \quad (4.4)$$

where
$$B_{\tilde{\nu}} = 2\tilde{\nu}^2 \, \frac{hc\tilde{\nu}}{\exp{(hc\tilde{\nu}/kT)} - 1} \quad (4.5)$$

has been written for the Planck function at wavenumber $\tilde{\nu}$ (cf. appendix 7). Throughout this chapter frequencies will be expressed in wavenumbers denoted by $\tilde{\nu}$, i.e. frequency in Hz divided by the velocity of light c. The quantities $I_{\tilde{\nu}}$ and $B_{\tilde{\nu}}$ are intensities of radiation at wavenumber $\tilde{\nu}$ per unit area per unit solid angle per unit wavenumber interval.

Of the quantities on the r.h.s. of (4.4) $B_{\tilde{\nu}}$ is a relatively slowly varying function of frequency whereas $\tau_{\tilde{\nu}}$ varies very rapidly with frequency (fig. 4.2). This is because molecules possess discrete energy levels associated with different vibrational and rotational states; any given vibration–rotation band contains many thousands of individual lines. The structure of the bands is often complex and is further complicated because each line in the band is broadened by collision broadening or Doppler broadening, broadening mechanisms which vary with temperature or pressure and are, therefore, considerably variable throughout the atmosphere. Given adequate information about the intensities, positions, widths and shapes of the spectral lines involved in any given spectral region, it is possible in principle to perform the integration over frequency in (4.4). However, to cope in practice with the integration it is necessary to find simplified methods of removing much of the complication associated with molecular band structure. We shall return to this problem later in the chapter (§ §4.4 and 4.8).

4.3 Heating rate due to radiative processes

The main reason why radiative transfer calculations are required is to deduce the contribution to the atmosphere's energy budget from radiative processes. The local heating rate is expressed in (2.4) in terms of the divergence of the upward and downward radiation fluxes.

For nearly all atmospheric situations it is adequate to employ the simple approximation of §2.2 for integration over angle, so that if in (4.4) τ is replaced by τ^* (i.e. pathlengths are multiplied by 5/3) and B is replaced by πB, an equation for the upward flux $F^\uparrow$ at level z_1 can be written

$$F^\uparrow = \int_0^\infty F_{0\tilde{v}}\, \tau_{\tilde{v}}^*(z_0, z_1)d\tilde{v} + \int_0^\infty \int_{\tau_{\tilde{v}}^*(z_0, z_1)}^1 \pi B_{\tilde{v}}(z)\, d\tau_{\tilde{v}}^*(z, z_1)\, d\tilde{v}$$

(4.6)

An entirely similar expression applies for the downward flux $F^\downarrow$.

4.4 Single lines

In §4.2 we saw that computation of fluxes and heating rates requires knowledge of the atmospheric transmission. Before attempting to compute the average transmission for a given spectral region, it is necessary to look in some detail at the structure of absorption bands.

The simplest model of a molecular band which may be employed is of a large number of lines which are independent (i.e. non-overlapping), broadened by collision processes only. For the carbon dioxide bands and water vapour bands of importance in the atmosphere this is a good approximation for levels in the stratosphere between $\sim$ 20 km and 60 km altitude. Above these levels Doppler broadening becomes important (problem 4.8) and at lower levels lines overlap very substantially.

Simple theory leads to an absorption coefficient $k_{\tilde{v}}$ at wavenumber $\tilde{v}$ due to a collision broadened line centred at $\tilde{v}_0$ given by (see for instance

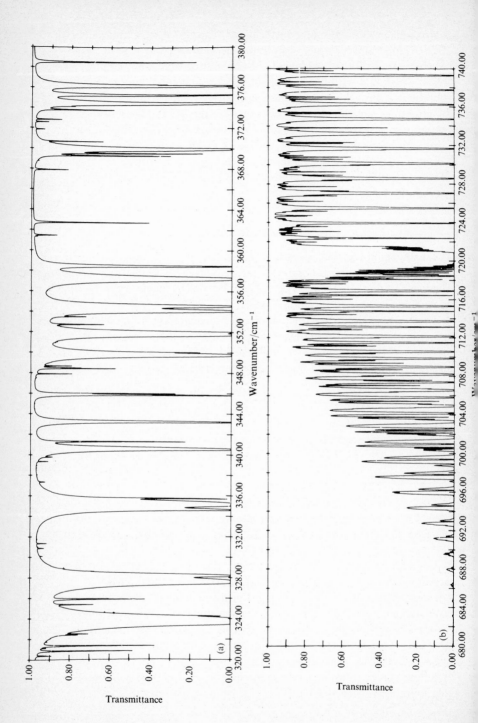

Houghton & Smith, 1966; Thorne, 1974)

$$k_{\tilde{\nu}} = \frac{s\gamma}{\pi\{(\tilde{\nu} - \tilde{\nu}_0)^2 + \gamma^2\}} \tag{4.7}$$

where
$$s = \int_0^\infty k_{\tilde{\nu}}\, d\tilde{\nu} \tag{4.8}$$

is the line strength and $\gamma = (2\pi tc)^{-1}$ the half-width (in wavenumbers). t is the mean time between collisions in the absorbing gas, a quantity which varies over several orders of magnitude in the atmosphere because it is inversely proportional to the pressure p. Ignoring the temperature dependence of γ which by comparison is small, we can write

$$\gamma = \gamma_0 p/p_0 \tag{4.9}$$

where γ_0 is its value at standard pressure p_0. For many gases, at STP, $\gamma_0 \simeq 0.1$ cm^{-1}, so that at 50 km altitude where the pressure is 1 mb, $\gamma \simeq 10^{-4}$ cm^{-1}.

Consider what appearance such a line would have when viewed with a high resolution laboratory spectrometer through a path of length l containing absorbing gas at density ρ. The transmission $\tau_{\tilde{\nu}}$ of such a path will be

$$\tau_{\tilde{\nu}} = \exp(-k_{\tilde{\nu}}\rho l) \tag{4.10}$$

and the *equivalent width* or *integrated absorptance* W of the line given by (fig. 4.3)

$$W = \int_{-\infty}^\infty d\tilde{\nu}(1 - \tau_{\tilde{\nu}})$$

$$= \int_{-\infty}^\infty d\tilde{\nu}\{1 - \exp(-k_{\tilde{\nu}}\rho l)\} \tag{4.11}$$

Two approximations for W are very useful: (1) the *weak* approximation where the exponential in (4.11) may be approximated by the first term of its expansion, in which case on substituting for $k_{\tilde{\nu}}$ from (4.7) (problem 4.2)

$$W = s\rho l \tag{4.12}$$

and (2) the *strong* approximation for which there is no transmission of any consequence near the line centre (the line centre is then said to be black). The

Fig. 4.2. Illustrating the fine structure of molecular absorption bands. Atmospheric transmission for a 10 km horizontal path at 12 km altitude as computed by McClatchey & Selby (1972) for (a) the 320–380 cm^{-1} (31–26 μm) region where the lines are due to water vapour, (b) the 680–740 cm^{-1} (15–13 μm) region where the lines are mainly due to carbon dioxide.

Only a very small part of the spectrum is covered in this diagram. Compare with fig. 2.1 in which the variation of atmospheric transmission with frequency is on a very much coarser scale.

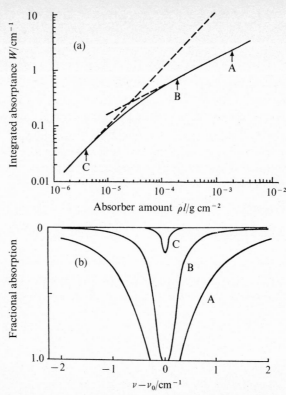

Fig. 4.3. (a) Curve of growth of a typical spectral line with $s = 10^4 \text{ cm}^{-1}(\text{g cm}^{-2})^{-1}$ and $\gamma_0 = 0.06 \text{ cm}^{-1}$ showing the linear and square root regions of growth. (b) Actual shapes of the spectral line for different values of ρl corresponding to the values shown by the arrows in (a).

actual value of $k_{\bar{\nu}}$ near the line centre is, therefore, not important providing it is large enough for γ^2 to be omitted from the denominator in (4.7). The integration in (4.11) may now be carried out (problem 4.3) and

$$W = 2(s\gamma\rho l)^{1/2} \tag{4.13}$$

Plotting the equivalent width W against the length of absorbing path l results in the *curve of growth* (fig. 4.3).

Taking in a substantial proportion of an absorption band of width $\Delta\bar{\nu}$ in which a number of non-overlapping lines are present, the average transmission $\bar{\tau}$ over the spectral interval will be

$$\bar{\tau} = 1 - \frac{\sum\limits_{i} W_i}{\Delta\bar{\nu}} \tag{4.14}$$

where W_i is the equivalent width of the ith line.

36

4.5 Transmission of an atmospheric path

We apply the results of the previous section to a vertical path in the atmosphere between levels of pressure p_1 and p_2 ($p_1 > p_2$). When the weak approximation applies this is straightforward as the equivalent width is then only a function of the amount of absorber in the path, i.e.

$$W = s \int_{\text{path}} c\rho \, dz \qquad (4.15)$$

where c is the fractional concentration (by mass) of absorber and ρ is the total density. Since $\rho \, dz = - dp/g$ and assuming a constant value of c along the path

$$W = sc(p_1 - p_2)/g \qquad (4.16)$$

It is not so clear how to proceed when the absorption is stronger, as the pressure (and hence γ) varies along the path. In this case application is made of the *Curtis–Godson approximation* which states that a mean pressure $\bar{p}$ may be defined for a path

$$\bar{p} = \frac{\int p c\rho \, dz}{\int c\rho \, dz} \qquad (4.17)$$

i.e. in deriving the mean pressure, the pressure along the path is weighted by the amount of absorber. This is an exact approximation when the strong approximation applies (problem 4.6); it is, of course, also exact under the weak approximation when the equivalent width is independent of pressure.

For uniform composition, $\bar{p} = \frac{1}{2}(p_1 + p_2)$ so that substituting from (4.9) and (4.17) in (4.13) we have in the strong approximation, for a single spectral line

$$W = 2\left\{ \frac{s\gamma_0 c}{2gp_0} (p_1^2 - p_2^2) \right\}^{1/2} \qquad (4.18)$$

in which case the average transmission over a spectral interval is, from (4.14) and (4.18)

$$\bar{\tau} = 1 - \frac{1}{\Delta\tilde{\nu}} \left\{ \frac{2c}{gp_0} (p_1^2 - p_2^2) \right\}^{1/2} \sum_i (s_i\gamma_{0i})^{1/2} \qquad (4.19)$$

where $\sum_i (s_i\gamma_{0i})^{1/2}$ is the sum of the square roots of the product of line strength and line width at STP within the spectral interval.

4.6 Cooling by carbon dioxide emission from upper stratosphere and lower mesosphere

We now apply the non-overlapping strong collision-broadened line approximation to the ν_2 band of carbon dioxide at 15 μm wavelength in the 20–60 km region. In this part of the atmosphere the temperature distribution is mainly determined by a balance between radiative cooling in the infrared from this carbon dioxide band and heating by absorption of solar radiation by ozone (§4.7).

To compute the radiation loss and hence the cooling rate we shall employ the *cooling to space* approximation, in which exchange of radiation between different atmospheric layers is neglected in comparison with the loss of radiation direct to space – a good approximation for this particular atmospheric region. Equating the loss of energy to space, the term $dF^\uparrow/dz$ from (2.4), to the cooling rate we have

$$\frac{dT}{dt}\rho c_p = - \int_{\Delta\tilde\nu} \pi B_{\tilde\nu}(T) \frac{d\tau_{\tilde\nu}^*(z,\infty)}{dz} d\tilde\nu \tag{4.20}$$

where the integration over frequency is as in (4.6) and covers the region $\Delta\tilde\nu$ of the 15 μm carbon dioxide band. Over this spectral interval $B_{\tilde\nu}(T)$ can be considered constant with frequency. From (4.19)

$$\int_{\Delta\tilde\nu} \tau_{\tilde\nu}^*(z,\infty)\, d\tilde\nu = \Delta\tilde\nu - p\left(\frac{10c}{3gp_0}\right)^{1/2}\sum_i (s_i\gamma_{0i})^{1/2} \tag{4.21}$$

The factor 5/3 is included because we require the slab transmission appropriate to flux calculations. The strong approximation has also been applied to all the lines in the band; the result of problem 4.14 shows that this is a valid simplification. Substituting in (4.20) and remembering that $\rho\, dz = -dp/g$ we have

$$\frac{dT}{dt} = -\frac{g\pi B_{\tilde\nu}(T)}{c_p}\left(\frac{10c}{3gp_0}\right)^{1/2}\sum_i (s_i\gamma_{0i})^{1/2} \tag{4.22}$$

Substituting values $c_p = 1005$ J kg^{-1}, $c = 4.7\times10^{-4}$, $\sum_i(s_i\gamma_{0i})^{1/2} =$ 1600 cm^{-1} (g cm^{-2})$^{-1/2}$ (appendix 10), we have

$$\frac{dT}{dt} = 2.0\pi B_{\tilde\nu} \ \text{K s}^{-1} \tag{4.23}$$

where $\pi B_{\tilde\nu}$ is in W cm^{-2} (cm^{-1})$^{-1}$. Note the important result that the cooling rate is independent of height.

4.7 Absorption of solar radiation by ozone

Ozone is formed in the stratosphere (10–50 km) and mesosphere (50–80 km) by photochemical processes (§5.5), the peak of ozone concentration occurring at about 25 km altitude. Ozone absorbs ultraviolet solar radiation in the Hartley band (200–300 nm); at levels below 70 km virtually all the energy absorbed goes into kinetic energy of the molecules, i.e. into increasing the atmospheric temperature.

The absorption spectrum of ozone in the Hartley band is mainly a continuous spectrum, with an absorption coefficient k_ν at frequency ν independent of pressure so that ignoring the complications of scattering which must be taken into account in an accurate calculation (cf. §12.5) the incident solar flux $F_{S\nu}(z)$ at height z is given by

$$F_{S\nu}(z) = F_{S\nu}(\infty) \exp\left(-\int_z^\infty k_\nu \rho_z \sec \theta \, dz\right) \tag{4.24}$$

where ρ_z is the density of ozone at height z, and θ is the zenith angle of the sun. The heating rate at level z is given by

$$c_p \rho \frac{dT}{dt} = \cos \theta \int_{\text{band}} \frac{dF_{S\nu}}{dz} \, d\nu \tag{4.25}$$

Given values of k_ν across the spectrum and ρ_z at various heights, numerical integration of (4.24) and (4.25) can be carried out (see appendices 8 and 9 for information about solar flux and ozone absorption). For a typical ozone distribution at mid latitudes, the peak of heating is at ~ 50 km altitude; when integrated over a day, it amounts to a rate of change of temperature of ~ 8 K day^{-1} (problem 4.9).

We can now equate this heating due to ozone absorption at 50 km altitude with cooling due to emission from carbon dioxide from §4.6. The equilibrium value of $\pi B_{\bar{\nu}}(T)$ at 50 km is found to be 4.6×10^{-5} W cm^{-2} (cm^{-1})$^{-1}$ equivalent to a temperature T of ~ 290 K, approximately as found at that level. A diurnal temperature variation of ~ 4 K is present at 50 km because heating only occurs during the day but cooling goes on all the time.

4.8 Band models

The simplest model of a band is that of non-overlapping lines described in §4.4. Such a model, however, is only applicable when the lines are well separated. For the atmosphere it can only be applied for the path lengths and pressures above altitudes of about 30 km. A variety of band models have been invented which allow for overlap, in particular the regular model first described by Elsasser, in which an absorption band is simulated by an array of equally spaced lines of the same shape and strength, and the random model due to Goody in which the lines are considered to be spaced randomly and to have a distribution of line strengths following some statistical law. This latter model leads to a particularly simple and useful expression for the average transmission $\bar{\tau}$ of a spectral interval of width $\Delta\bar{\nu}$ containing a large number of lines

$$\bar{\tau} = \exp\left(-\Sigma W_i / \Delta\bar{\nu}\right) \tag{4.26}$$

where ΣW_i is the sum of the equivalent widths of all the lines in the interval considered as independent lines. In problem 4.13 the theory of the random band model is developed further. From this theory and from the spectral data provided in appendix 10, the transmission of any spectral interval in the bands of water vapour, carbon dioxide and ozone in the region of the thermal infrared can be calculated.

39

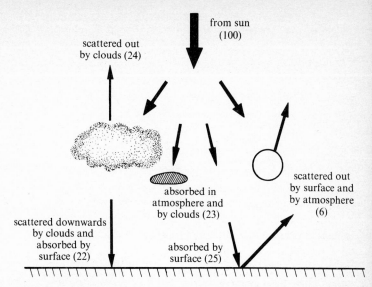

Fig. 4.4 The interactions of incident solar radiation with the earth and its atmosphere.

With the use of band models, the frequency integration of (4.4) may be carried out for rather few spectral intervals, each major absorption band being divided into just one or two spectral regions. Such simplified radiation calculations are required for incorporation into numerical models (§11.7).

4.9 Continuum absorption

In the wavelength region 8–13 μm in the infrared, apart from the ozone band at 9.6 μm, the atmosphere is reasonably transparent (fig. 2.1). Some absorption by water vapour lines is still present, however, and of more importance, by water vapour dimers, i.e. associations of water vapour molecules in pairs. This latter absorption is a continuum with absorption coefficient proportional to the partial pressure of water vapour; associations are more likely to occur when the partial pressure is high. For this reason this continuum absorption is particularly important in humid atmospheres; for a wet tropical atmosphere, transmission of a vertical path through the atmosphere may only be ~ 50% in this region (problem 4.11).

4.10 Global radiation budget

In this chapter, equations have been written down which, given detailed information regarding the spectra of the molecules, enable calculations of the radiation fluxes and the radiation heating rate to be made at any level in a clear atmosphere. No account has been taken of clouds and their radiative

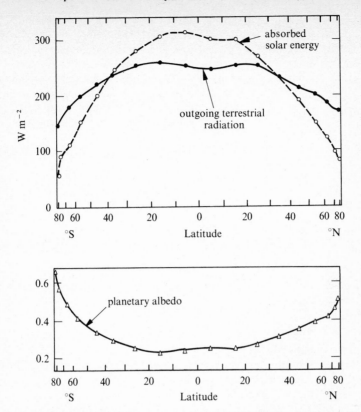

Fig. 4.5. Average components of the earth's radiation budget as deduced from satellite observations, 1962–66, by Vonder Haar &
Suomi (1971).

properties. Clouds are, in fact, probably the dominant influence in the radiative budget of the lower atmosphere but adequately taking them into
account raises many problems, which will be considered briefly in §6.4.

Of fundamental importance to our understanding of the atmosphere's
energy budget is adequate knowledge of the components of the radiation
budget at the top of the atmosphere. Fig. 4.4 shows what happens on average
to the incident solar radiation. The outgoing long-wave radiation originates
from near the surface or from cloud top in the window region and from high
in the troposphere in the strong carbon dioxide and water vapour bands (cf.
fig. 12.3). Fig. 4.5 shows the results of satellite measurements of the two
main components of the radiation budget averaged over longitude and over a
complete year. The excess of solar energy absorbed over that emitted in equatorial regions must be transported through atmospheric motions or ocean currents to the polar regions where it is required to supply the deficit between
energy emitted and received. How atmospheric motions achieve the transfer
will be discussed in chapter 10.

41

More complex radiation transfer

Problems

4.1   Integrate (2.3) to derive (4.3) using the integrating factor $\exp(-\chi)$

4.2   Integrate (4.11) under the weak approximation to obtain (4.12)

4.3   Integrate (4.11) under the strong approximation to obtain (4.13)

4.4   Consider a path of fixed length l in which the pressure of an absorbing gas can be varied. Show that under conditions of collision broadening the absorption at the line centre is independent of pressure.

4.5   Show that for a single collision broadened line the integral of (4.11) may be written as

$$W = \int_{-\infty}^{\infty} dx \left\{ 1 - \exp\left(\frac{-2u\gamma^2}{x^2 + \gamma^2}\right)\right\}$$

where $x = \tilde{\nu} - \tilde{\nu}_0, \qquad u = s\rho l / 2\pi\gamma$

The value of this integral may be expressed in terms of the Bessel function $J(u)$ of the first kind with imaginary arguments giving

$$W = 2\pi\gamma l u \exp(-u)\{J_0(u) + J_1(u)\}$$

$$= 2\pi\gamma L(u) \tag{4.27}$$

The function $L(u)$ is known as the *Ladenberg and Reiche function*.

4.6   Substitute the strong approximation to $k_{\tilde{\nu}}$ from (4.7) (i.e. omit the γ^2 in the denominator) into (4.1) for the transmission along an atmospheric path and show that the Curtis–Godson approximation (4.17) is exact in the case of the strong approximation.

4.7   Two useful approximations to the equivalent width W of a collision broadened line are

$$W = s\rho l \left(1 + \frac{s\rho l}{4\gamma}\right)^{-1/2}$$

and

$$W = s\rho l \left\{1 + \left(\frac{s\rho l}{4\gamma}\right)^{5/4}\right\}^{-2/5}$$

Show that these expressions have the correct 'weak' and 'strong' limits ((4.12) and (4.13)). The first deviates by less than 8% from the correct equivalent width for all values of the parameters. The second due to Goldman (1968) has a maximum error of about 1%.

4.8   The shape of a spectral line due to Doppler broadening is given by

$$k_{\tilde{\nu}} = \frac{s}{\gamma_D \pi^{1/2}} \exp\left\{-\left(\frac{\tilde{\nu} - \tilde{\nu}_0}{\gamma_D}\right)^2\right\} \tag{4.28}$$

where $\gamma_D = \frac{\tilde{\nu}_0}{c}\left(\frac{2RT}{M_r}\right)^{1/2}$ \hfill (4.29)

(M_r is molecular weight).

Calculate the pressure in the atmosphere at which the half-width due to collision broadening is equal to that due to Doppler broadening for (1) a carbon dioxide line at $667 \, \text{cm}^{-1}$, (2) a water vapour line at $1600 \, \text{cm}^{-1}$, (3) a water vapour line at $100 \, \text{cm}^{-1}$. Assume all lines have collision broadened half-widths of $0.1 \, \text{cm}^{-1}$ at standard pressure.

For the two shapes compare the ratio of the absorption coefficient at the line centre to its value several half-widths away. Notice that the absorption coefficient in the wings of collision broadened lines is relatively much greater than that for the wings of Doppler broadened lines. For this reason, even for conditions where the Doppler half-width is greater than the collision broadened half-width the transfer of radiation in the collision broadened wings of the lines remains important.

4.9   Assuming the sun overhead and a uniform temperature atmosphere, for a distribution of the number density n_3 of ozone molecules as a function of pressure p given by $n_3 = n_0 p^{3/2}$ (quite a good approximation to the upper part of the ozone layer), show that the heating rate h by absorption of solar radiation is of the form

$$\frac{h}{h_m} = \left(\frac{p}{p_m}\right)^{1/2} \exp\left\{-\frac{1}{3}\left(\frac{p}{p_m}\right)^{3/2} - 1\right\} \qquad (4.30)$$

where p_m is the level of maximum heating h_m. In your calculation assume a single absorption coefficient independent of wavelength within the ozone band.

Given that the level of maximum heating is at 1 mb pressure, the solar flux over the ozone absorbing region is $0.7 \times 10^{-3} \, \text{W cm}^{-2}$, calculate h_m.

4.10   Because of the ellipticity of the earth's orbit the solar radiation incident on the earth is about 7% greater in December than in June. Assuming constant ozone, from the variation with temperature of the Planck function at $15 \, \mu\text{m}$, estimate the temperature difference between the temperature of the stratopause (the level of maximum temperature near 50 km) in the summer over the north pole and over the south pole. Compare your result with the observed temperature difference of about 2.5 K (Barnett (1974)).

4.11   Absorption in the atmospheric window between $8 \, \mu\text{m}$ and $13 \, \mu\text{m}$ is mostly due to the water vapour dimer; the absorption coefficient is of the form $k_2 e$ where e is the water vapour pressure (in mb), $k_2 \cong 10^{-2} \, (\text{g cm}^{-2})^{-1} \, \text{mb}^{-1}$. If the water vapour pressure near the surface is 10 mb, calculate (1) the transmission of a horizontal path 1 km long near the surface, (2) the transmission of a vertical path of atmosphere assuming that the distribution of water vapour pressure is proportional to (pressure in atmospheres)[4]. Also estimate for a layer near the surface the cooling rate in K day^{-1} by radiation from water vapour in this spectral region.

4.12 Integrate (4.3) by parts to give

$$I_1 = \{I_0 - B(z_0)\}\tau(z_0,z_1) + B(z_1) + \int_{z_1}^{z_0} \tau(z,z_1) \frac{dB(z)}{dz} dz$$

Write this equation in terms of flux and using (2.4) show that when $I_0 = B(z_0)$ an expression for the atmospheric heating rate at level z_1 is

$$\rho c_p \frac{dT}{dt} = \pi \int_0^\infty \int_0^\infty \frac{d\tau_{\tilde{\nu}}^*(z,z_1)}{dz_1} \cdot \frac{dB_{\tilde{\nu}}(z)}{dz} dz \, d\tilde{\nu}$$

4.13 Suppose in a random array of spectral lines the number $N(s)/ds$ of lines having a strength between s and $s + ds$ is

$$N(s)ds = \frac{N_0}{s} \exp\left(-\frac{s}{\sigma}\right) \tag{4.31}$$

This distribution is known as the Malkmus model (Malkmus, 1967) and fits well to line distributions found in practice. Show that the sum of the equivalent widths ΣW_i of these lines (assuming no overlap) after passing through a path of length l with absorber density ρ is

$$\Sigma W_i = \int_0^\infty \frac{N_0}{s} \exp\left(-\frac{s}{\sigma}\right) ds \int_{-\infty}^\infty \left\{1 - \exp\left(-k_{\tilde{\nu}}\rho l\right)\right\} d\tilde{\nu} \tag{4.32}$$

If a line shape is described by

$$k_{\tilde{\nu}} = sf(\tilde{\nu}) \tag{4.33}$$

carry out the integration over s to give

$$\Sigma W_i = N_0 \int_{-\infty}^\infty \ln\{1 + \sigma\rho l f(\tilde{\nu})\} d\tilde{\nu} \tag{4.34}$$

If $k_{\tilde{\nu}}$ is given by the collision broadened expression (4.7) show that

$$\Sigma W_i = 2\pi\gamma N_0 \left\{ \left(1 + \frac{\sigma\rho l}{\pi\gamma}\right)^{1/2} - 1 \right\} \tag{4.35}$$

Compare the values of ΣW_i given by (4.35) in the limits of $\sigma\rho l/\pi\gamma \ll 1$ and $\gg 1$ with what would be expected from the weak and strong approximations respectively (4.12) and (4.13)

$$\Sigma_i W_i \text{ (weak)} = \Sigma_i s_i\rho l \tag{4.36}$$

$$\Sigma_i W_i \text{ (strong)} = 2 \Sigma_i (s_i\gamma_i\rho l)^{1/2} \tag{4.37}$$

where the summation in each case is over all the lines denoted by strength s_i and half-width γ_i.

The best way of fitting spectral line data to the random model is not by attempting to match the model distribution of line strengths with the actual distribution but rather by matching the weak and strong limits of (4.35) with (4.36) and (4.37) respectively. By doing this, find expressions for $\pi N_0 \gamma$ and $\sigma/\pi\gamma$ in (4.35) in terms of Σs_i and $\Sigma(s_i\gamma_i)^{1/2}$. Hence show that (4.35) may be written

$$\Sigma\ W_i = \frac{2\{\Sigma(s_i\gamma_i)^{1/2}\}^2}{\Sigma s_i}\left[\left\{1+\frac{\rho l(\Sigma s_i)^2}{[\Sigma(s_i\gamma_i)^{1/2}]^2}\right\}^{1/2}-1\right] \qquad (4.38)$$

This is a particularly useful form of $\Sigma\ W_i$ which, for any given spectral interval $\Delta\tilde{\nu}$ can be substituted in (4.26) to calculate the transmission under any given circumstance when collision broadening applies. Spectral line data for many bands of importance suitable for substituting in (4.38) are listed in appendix 10.

4.14   For calculations of the transmission of a vertical path of atmosphere between a level where the pressure is p and the top of the atmosphere, show, by applying the data of appendix 10 for the $15\ \mu m$ carbon dioxide band to (4.38) that, providing collision broadening applies, the strong approximation is a very good assumption.

5

The upper atmosphere

5.1 Upper atmospheric temperature structure

In fig. 5.1 is shown a typical cross-section of the temperature structure of the earth's atmosphere from the surface up to ~ 100 km. The temperature falls with increasing height in the lowest 10 or 15 km roughly at the adiabatic lapse rate as discussed in chapter 2. Above the tropopause the main features of the temperature structure are determined by radiative processes. The high temperature at ~ 50 km (known as the *stratopause*) is due to absorption of solar radiation by ozone (§4.7) – balanced mainly by emission of infrared radiation by carbon dioxide. Above the temperature peak at 50 km is the *mesosphere* (middle sphere) where the temperature falls with height although not so steeply as in the troposphere. The upper boundary of the mesosphere is the temperature minimum around 80 km altitude known as the *mesopause* where there is rather little absorption of solar radiation. Above the mesopause, solar ultraviolet radiation is strongly absorbed particularly by molecular and atomic oxygen and the temperature rises rapidly to between 500 K and 2000 K in the region known as the *thermosphere*.

The variation of temperature with latitude should be noted from fig. 5.1. At the earth's surface and at the level of the stratopause the equator is warmer than the polar regions as simple considerations would predict. The tropopause and the mesopause are, however, substantially colder over the equator than over the poles. This is particularly surprising for the winter polar regions where there is no solar radiation. Radiative considerations cannot provide an explanation of these reversed temperature gradients. Their existence demonstrates that atmospheric motions not only act as heat engines and transfer energy from source regions to sink regions but can also act as refrigerators transferring energy in the other direction.

In the following sections of this chapter we shall consider various physical processes which occur in the atmosphere above 30 km. Only 1% of the atmospheric mass lies above 30 km and only 1 part in 10^5 above ~ 80 km. It must not be expected, therefore, that events in these upper regions have a very immediate or large effect on the structure or motion of the troposphere. However, because of the close links which exist between motions in various parts of the atmosphere, study of upper atmospheric motion can have an important bearing on our understanding of motion lower down. Further, what happens at the top of the atmosphere is important in the study of the long-term evolution of the atmosphere. Consideration, therefore, of upper atmosphere phenomena, in addition to being of interest in its own right, plays an important part in the understanding of the whole.

5.2 Diffusive separation

The amount of mixing in the lower atmosphere ensures that no significant diffusive separation occurs between the heavy atmospheric constituents (e.g. argon) and the light ones (e.g. hydrogen). For practical purposes, therefore, the main constituents of the lower atmosphere are uniformly mixed apart from those which are involved in phase changes (e.g. water vapour) or chemical changes (e.g. ozone).

Although turbulent mixing dominates over molecular diffusion in the lower atmosphere, because the molecular diffusion coefficient is proportional to the mean free path and hence inversely proportional to density, in the very high atmosphere molecular diffusion plays the dominant role both for the transfer of the molecules themselves and for the transfer of heat. A typical value of molecular diffusion coefficient for an atmospheric gas is $\sim 10^{-4} \, m^2 s^{-1}$ at STP and therefore $\sim 100 \, m^2 s^{-1}$ at a pressure of 10^{-6} atm (corresponding to ~ 96 km altitude). This is of the same order as a typical eddy diffusion coefficient (cf. §9.2) appropriate to the lower thermosphere. The lower boundary of the region above which molecular diffusion dominates is ~ 120 km altitude and is known as the *turbopause* (sometimes known as the *homopause*: the region above, where molecular diffusion is dominant, is sometimes known as the *homosphere*). The turbopause can often be observed fairly clearly from the luminous trails of rockets, as a boundary below which the trail is violently disturbed by turbulence and above which it is relatively smooth.

We first consider the composition of an atmosphere in a state of diffusive equilibrium. In such equilibrium the hydrostatic equation (1.4) is applied to each constituent separately. This follows from Dalton's law of partial pressures, together with the condition that there be no net transport of any constituent across a horizontal surface. For the case of an isothermal atmosphere

$$\ln \{p(i)/p_0(i)\} = -z/H(i) \tag{5.1}$$

where the index i refers to the ith constituent. If $T = 1000$ K, the order of temperature found in the thermosphere, the values of the scale height H for argon ($M_r = 40$), nitrogen ($M_r = 28$), atomic oxygen ($M_r = 16$) and atomic hydrogen ($M_r = 1$) are respectively 21 km, 30 km, 54 km and 850 km. From (5.1), the ratio of the number densities n of two different constituents at an altitude z may be compared to their number densities n_0 at altitude z_0 at which diffusive separation begins. For example, for argon and nitrogen

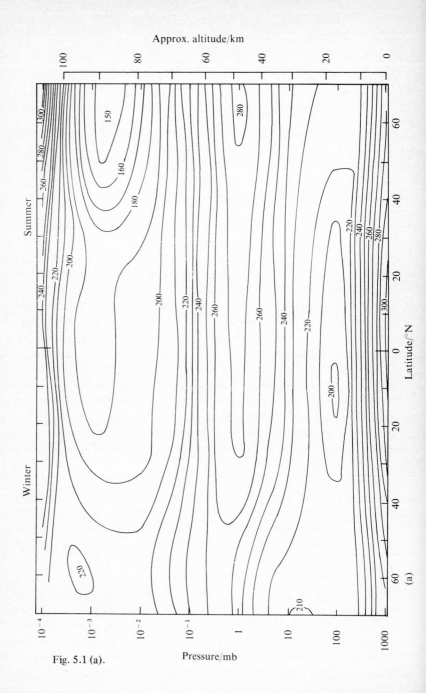

Fig. 5.1 (a).

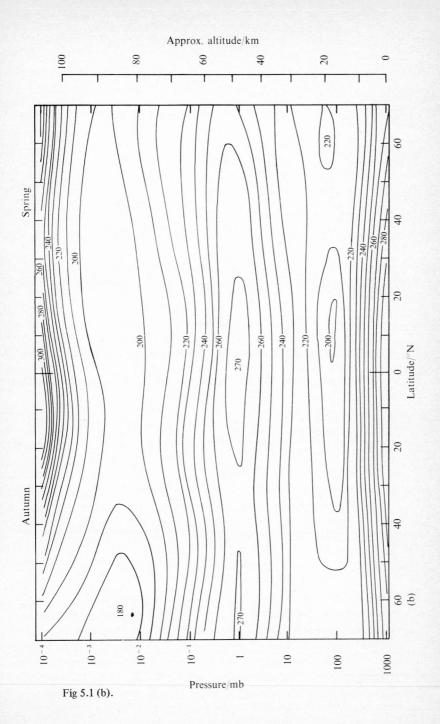

Fig 5.1 (b).

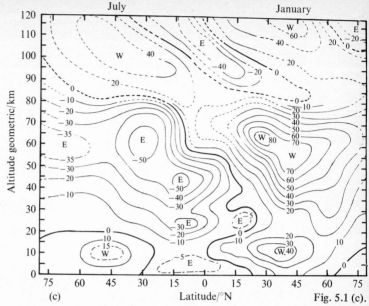

Fig. 5.1. Mean latitudinal cross-sections of temperature (K), surface to ~ 100 km, for northern hemisphere for (a) summer and winter, (b) spring and autumn. Data for the diagrams tabulated in appendix 5 where sources of data are listed. (c) mean January and July zonal winds (from COSPAR, 1972). The relation between the temperature and wind fields is described in §7.6.

$$\frac{n(A)n_0(N_2)}{n_0(A)n(N_2)} = \exp\left[-(z-z_0)\left\{\frac{1}{H(A)} - \frac{1}{H(N_2)}\right\}\right] \tag{5.2}$$

The results of such calculations are shown in fig. 5.2. Whereas nitrogen is the most abundant consitituent in the lower atmosphere, this role is taken over by atomic oxygen at about 170 km and by helium at about 500 km, these altitudes being quoted for mean atmospheric conditions; they vary very considerably with atmospheric temperature (problem 5.1).

5.3 The escape of hydrogen

As density decreases in the atmosphere with increasing altitude, eventually a level is reached where collisions are so rare that a molecule moving upwards will turn about under the influence of gravity and return again to the denser layers without making a collision. Under such conditions the faster molecules escape from the atmosphere altogether. This region is called the *exosphere*. The *critical level* z_c for such escape may be defined such that a proportion e^{-1} of a group of very fast particles moving vertically upwards at z_c will experience no collisions as they pass completely out of the atmosphere. For

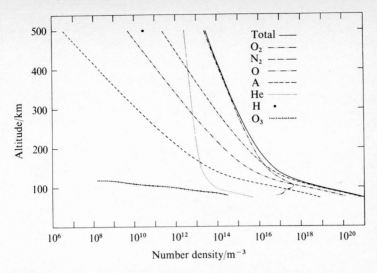

Fig. 5.2. Total number density and number densities of various constituents for mean atmospheres (from COSPAR, 1972).

molecules of diameter d, the probability of no collision being made in a vertical path dz is $\exp\left[-\pi d^2 n(z)\,dz\right]$, where $n(z)$ is the particle density at height z. Substituting for $n(z)$ from the hydrostatic equation (1.4) the condition for the critical level z_c is, therefore

$$\int_{z_c}^{\infty} \pi d^2\, n(z_c)\, \exp\left[\frac{-(z-z_c)}{H}\right] dz \,=\, 1 \tag{5.3}$$

i.e. $\qquad n(z_c) \,=\, (\pi d^2 H)^{-1} \tag{5.4}$

The departure from equilibrium is sufficiently small for the scale height H to be considered approximately constant through the region. Since $[n(z)\pi d^2]^{-1}$ is the mean free path in the horizontal direction at level z, the critical level may also be defined as that at which the mean free path in the horizontal direction is equal to the scale height H. The critical level occurs at an altitude where the most abundant constituent is atomic oxygen (fig. 5.2). For a temperature of 1000 K (approximate value for mean solar activity (fig. 5.3)) its scale height H is 53 km. With $d \doteq 2 \times 10^{-10}$ m, $n(z_c) \simeq 10^{14}$ m^{-3} corresponding to a height z_c of between 400 and 500 km (fig. 5.2). Considerable variations, of course, occur with solar activity, time of day and season.

We now consider the rate of escape of hydrogen atoms. These originate mainly from molecules of water vapour and methane which are dissociated in the stratosphere and mesosphere by the action of ultraviolet radiation from the sun. For escape upwards from the earth's gravitational field a particle of mass m and velocity V must possess a kinetic energy $\tfrac{1}{2}mV^2$ greater than its gravitational potential energy mGM/a (G is the gravitational constant and M

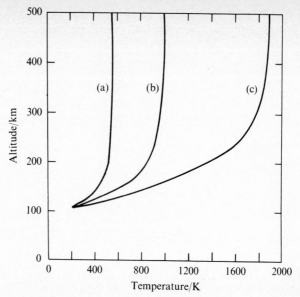

Fig. 5.3. Mean temperature of the upper atmosphere during periods of (a) very low, (b) mean, (c) very high, solar activity (from COSPAR, 1972). An indication of variations with time of day is given in fig. 5.4.

and a respectively the mass and radius of the earth): i.e.

$$\tfrac{1}{2}mV^2 > \frac{mGM}{a}$$

or $\qquad V^2 > 2ga, \quad \text{since } g = \frac{GM}{a^2}$ $\qquad\qquad$ (5.5)

or $\qquad V > V_c$

where $\quad V_c = 11 \, \text{km s}^{-1}$

This is over twice the mean molecular velocity of atomic hydrogen at a temperature of 1000 K, and very much larger than the mean velocities of other molecules.

Since the rate of escape is small the velocity distribution at the escape level will be approximately Boltzmann. The number dn_H of hydrogen atoms per unit volume having velocities between V and $V + dV$ is (see for instance Jeans, 1940)

$$dn_H = \frac{n_H(z_c)V^2 \, dV \exp(-\alpha V^2)}{\int_0^\infty V^2 \, dV \exp(-\alpha V^2)}$$ $\qquad\qquad$ (5.6)

where $\alpha = m/2kT$ and m is the mass of a hydrogen atom. Of these the number moving in an upward direction across unit area per second is $\tfrac{1}{4}V \, dn_H$

52

(see for instance Jeans, 1940), so that the number $\dot{N}$ of atoms escaping per unit time from unit area is

$$\dot{N} = \tfrac{1}{4}n_H(z_c) \left(\frac{\alpha^3}{\pi}\right)^{1/2} \int_{V_c}^{\infty} V^3 dV \exp\left(-\alpha V^2\right) dV \tag{5.7}$$

where a further factor $\tfrac{1}{4}$ has been included to allow for the fact that the number density at the critical level was computed for travel vertically upwards whereas (5.7) includes molecules travelling in any upward direction. The integral in (5.7) may be evaluated by parts to give

$$\dot{N} = \tfrac{1}{8}n_H(z_c) \left(\frac{\alpha}{\pi}\right)^{1/2} \left(V_c^2 + \frac{1}{\alpha}\right) \exp\left(-\alpha V_c^2\right) \tag{5.8}$$

$$= \beta\, n_H(z_c), \text{ say} \tag{5.9}$$

where β has the dimensions of a velocity.

For hydrogen and $T = 1000\,\mathrm{K}$, $\beta \simeq 1.6\,\mathrm{m\,s^{-1}}$.

The flux $\dot{N}$ of hydrogen atoms has, of course, to be maintained right through the atmosphere up to the escape level. Above the turbopause, it is maintained predominantly by molecular diffusion, the net flux of atoms at any level due to diffusion being (omitting the rather small thermal diffusion term)

$$D\left\{\frac{dn_H}{dz} - \left(\frac{dn_H}{dz}\right)_e\right\} \tag{5.10}$$

where D is the diffusion coefficient and $(dn_H/dz)_e$ the vertical gradient of n_H under equilibrium conditions which from the hydrostatic equation is $-n_H/H_H$, H_H being the scale height for hydrogen atoms. Carrying through the calculation (problem 5.5) shows that to provide the required upward flux of hydrogen atoms, the ratio of hydrogen atom concentration at $120\,\mathrm{km}$ to that at the escape level must be substantially greater than it would be if no escape occurred.

5.4 The energy balance of the thermosphere

Above the turbopause at $\sim 120\,\mathrm{km}$ altitude, we have already seen in §5.2 that, because of the low density and the positive temperature gradient with altitude, molecular diffusion is more important than eddy diffusion. The same is true for heat transfer. The dominant processes controlling the temperature of the thermosphere are the absorption of solar radiation in the extreme ultraviolet (wavelengths below $91\,\mathrm{nm}$) by atomic oxygen, balanced by molecular conduction of heat. Of lesser importance under normal conditions are radiative transfer by far infrared transitions within the ground state of atomic oxygen and transfer of energy from particles which make up the *solar wind*. Under disturbed conditions the latter can become the dominant source of energy.

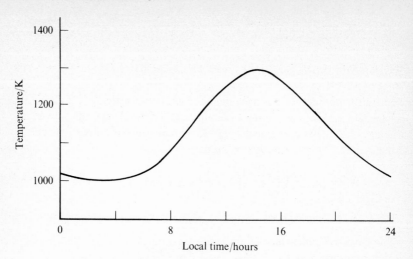

Fig. 5.4. The variation over the equator at the equinox of exospheric temperature with local time for a case when the minimum temperature is 1000 K (from COSPAR, 1972).

Ignoring, for the moment, variations through the day, consider equilibrium conditions for the atmospheric layer above level z. Since the conduction of heat across the top of the atmosphere is zero the solar radiation absorbed within a layer averaged over a day must be equal to the conduction of heat out of the bottom of the layer, i.e.

$$\epsilon\{\bar{I}(\infty) - \bar{I}(z)\} = \lambda(z)\frac{dT(z)}{dz} \qquad (5.11)$$

where $\bar{I}(z)$ is the average incident solar radiation, $\lambda(z)$ the coefficient of thermal conductivity and $T(z)$ the temperature, all at level z. The factor ϵ allows for the fact that not all the energy absorbed will be transformed into heat; some will be re-radiated out of the region altogether and some of the ionized or dissociated particles will diffuse out of the region to recombine elsewhere. According to simple kinetic theory, λ is proportional to the mean molecular velocity $\bar{c}$ which varies as $T^{1/2}$ but is independent of the pressure.

Knowing the distribution of atomic oxygen (§5.2) and its ultraviolet properties and also making some assumption about the value of ϵ, (5.11) may be solved numerically (problem 5.7). Fig. 5.3 shows the result of such calculations for different solar activities.

The problem of thermospheric heat balance may also be written in time-dependent form so that the diurnal variation may be estimated. Fig. 5.4 shows the variation of thermospheric temperature through the day as derived from such a calculation.

54

5.5 Photochemical processes

We have already seen the importance of ozone in determining the temperature structure of the stratosphere and mesosphere. The most important photochemical processes in the atmosphere, therefore, are those which lead to the formation and destruction of ozone. The simple 'classical' theory of ozone due to Chapman (1930) considered reactions involving oxygen only, namely:

$$O_2 + h\nu \rightarrow O + O \quad (J_2) \tag{5.12}$$

$$O + O + M \rightarrow O_2 + M \quad (k_1) \tag{5.13}$$

$$O + O_2 + M \rightarrow O_3 + M \quad (k_2) \tag{5.14}$$

$$O + O_3 \rightarrow 2O_2 \quad (k_3) \tag{5.15}$$

$$O_3 + h\nu \rightarrow O_2 + O \quad (J_3) \tag{5.16}$$

Equations (5.12) and (5.16) describe photodissociation in the presence of solar radiation, the first occurring for wavelengths less than 246 nm and the second for wavelengths less than 1140 nm but most strongly below 310 nm. J_2 and J_3 are dissociation rates per molecule per second, which will, of course, vary with altitude. k_1, k_2 and k_3 are reaction rates defined such that for reactions between two species present in number densities n_1 and n_2 respectively per unit volume, the number of reactions per second will be the product $n_1 n_2$ multiplied by the reaction rate. Reactions (5.13) and (5.14) are three-body

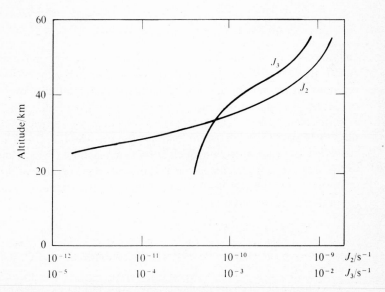

Fig. 5.5. Values of dissociation rates J_2 and J_3 averaged over a day for typical mid-latitude conditions (from Crutzen, 1971).

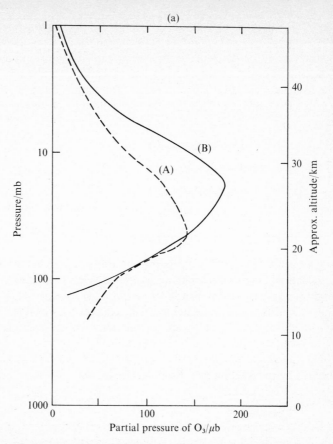

collisions, the third body M being required to satisfy energy and momentum conservation simultaneously. For these reactions to obtain the number of reactions per second, the appropriate reaction rate is multiplied by the product of the number densities of the three species involved.

In the stratosphere reaction (5.13) is slow and may be neglected. Reactions (5.14) and (5.16), in which 'odd' oxygen particles (i.e. O and O_3) are converted into each other, are much faster than reactions (5.12) and (5.15) in which 'odd' oxygen particles are created or destroyed. Because of this, in the stratosphere, equilibrium between the O and O_3 concentration is governed by (from the law of Mass Action)

$$J_3 n_3 = k_2 n_2 n_1 n_M \tag{5.17}$$

where n_1, n_2, n_3 and n_M are respectively the concentrations of O, O_2, O_3 and all molecules.

The balance between creation and destruction of odd oxygen particles is described by

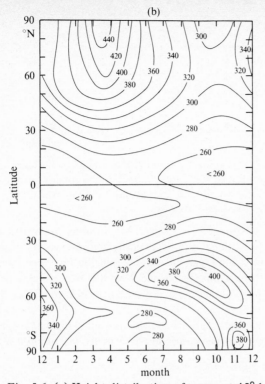

Fig. 5.6. (a) Height distribution of ozone at 45° latitude, summer (A) as observed and (B) as calculated from 'classical' pure oxygen photochemical theory (after Dütsch, 1968). (b) Total ozone in a vertical column (expressed as the depth of the column of gas if isolated from the rest of the atmosphere and at STP in units of 10^{-5} m, sometimes known as Dobson units) plotted as a function of latitude and season (after Dütsch, 1971). Note the maxima in spring at high latitudes in both hemispheres which result from transport by atmospheric motions.

$$2J_2 n_2 = 2k_3 n_1 n_3 \qquad (5.18)$$

From (5.17) and (5.18) the equilibrium ozone concentration is:

$$n_3 = n_2 \left(\frac{J_2 k_2 n_M}{J_3 k_3}\right)^{1/2} \qquad (5.19)$$

The rates J_2 and J_3 depend differently on altitude (fig. 5.5), J_2 dropping off with decreasing altitude more rapidly than J_3, mainly because of the overlap of ozone and oxygen absorption in the 200 nm region. A peak of ozone concentration, therefore, occurs (fig. 5.6). Agreement between the computed ozone profile and that observed is not very good, because (1) other reactions

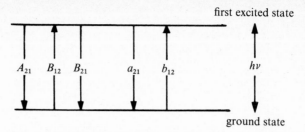

first excited state

ground state

Fig. 5.7. A two level system with Einstein coefficients: A_{21} for spontaneous emission, B_{12} for absorption, B_{21} for stimulated emission, a_{21} and b_{12} are probabilities per unit time of relaxation or excitation by collisional processes.

involving minor constituents such as NO, NO_2 or other particles such as H, OH, HO_2 themselves formed photochemically are involved in the formation and destruction of O and O_3 and because (2) the lifetime of ozone molecules at levels below 30 km is long (problem 5.9) so that ozone is redistributed by atmospheric motions. In fact the ozone mixing ratio is an important tracer of these motions (fig. 5.6).

Some photochemical reactions give rise to excited molecular species. The resulting radiation is known as *airglow*. A particularly important reaction is

$$H + O_3 \rightarrow OH^* + O_2 \tag{5.20}$$

which occurs near the mesopause (~ 90 km) and leads to vibrationally excited OH molecules which radiate in the near infrared. The amount of emission is large and makes a significant contribution to the energy budget of the mesopause region (problem 5.10).

5.6 Breakdown of thermodynamic equilibrium

In the discussion of radiative transfer in chapters 2 and 4 the assumption of local thermodynamic equilibrium (LTE) enabled Kirchhoff's law to be applied so that the Planck black-body function could be employed as the source function in the equation of transfer (2.3). At high levels in the atmosphere LTE is no longer a good assumption, and the molecular processes involved need to be considered in more detail.

Consider a system (fig. 5.7) of the ground state and first excited state of a particular vibrational mode. At atmospheric temperatures very few molecules will be in states higher than the first and we shall neglect these higher states in this simplified treatment. Molecules can be excited into the upper level either by absorbing radiation of the appropriate frequency or through the effect of collisions. Molecules can lose their vibrational energy and transfer back to the ground state by emitting radiation either spontaneously or by stimulated emission or again through collisions. If the volume of gas under

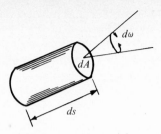

Fig. 5.8.

consideration is completely enclosed so that there is no net gain or loss by radiation a Boltzmann balance between the rate of excitation and de-excitation will be provided through the radiation and collision processes considered separately, and the ratio of the population of the upper level to that of the lower level will be the Boltzmann factor $\exp(-h\nu/kT)$. If now part of the enclosure is removed so that some net radiation is gained or lost from the system a Boltzmann balance cannot be maintained through absorption and emission processes. However, it is still possible that the rates of excitation and de-excitation by collision will be sufficiently rapid that they dominate over the radiation processes; the population of the upper level will then be affected only a little by the radiative gain or loss. This is the situation of local thermodynamic equilibrium. We may expect it no longer to apply when (1) there is a significant net radiative gain or loss from the excited level in question and when (2) the rates of excitation or de-excitation by collisions are comparable with the rates of excitation or de-excitation by absorption or emission.

To set up the problem, the first treatment of which was by Milne (1930), we need to express parameters describing the radiation field such as the absorption coefficient and the source function in terms of the detailed molecular rate constants of fig. 5.7.

First considering emission, the radiation power emitted by the element of volume dv (fig. 5.8) containing $n\,dv$ molecules, between frequencies ν and $\nu + d\nu$, and in solid angle $d\omega$ is $k_\nu n J_\nu\,ds\,dA\,d\nu\,d\omega$ where k_ν is the absorption cross-section per molecule – an expression which serves to define the source function J_ν. Assuming that J_ν is isotropic and, over the spectral region occupied by the absorption band, does not vary with ν, since $\int_{\Delta\nu} k_\nu\,d\nu = S$ (the total band strength of the vibrational transition, i.e. integrated over all rotational structure within the whole band of width $\Delta\nu$), the total power emitted by the element is $4\pi S n J_\nu\,dv$.

Turning this expression for total power emitted into a number of photons per unit time enables it to be related to the Einstein coefficient A_{21} (fig. 5.7),

i.e. $\qquad 4\pi S n J_\nu (h\nu)^{-1} = A_{21} n_2$ $\hfill (5.21)$

Note that in (5.21) we have neglected stimulated emission compared with spontaneous emission as for the bands of interest under atmospheric conditions spontaneous emission dominates.

In a similar way absorption of radiation in the element of volume dv can be considered. If I_ν is the incident intensity within the solid angle $d\omega$ and the spectral interval $d\nu$, the power absorbed will be $k_\nu I_\nu n \, dv \, d\nu \, d\omega$ which after integration gives for the total power absorbed

$$4\pi Sn \, d\nu \, \overline{I_\nu}$$

where

$$\overline{I_\nu} = \frac{1}{4\pi S} \int_{\Delta\nu} \int_{4\pi} k_\nu I_\nu \, d\omega \, d\nu \tag{5.22}$$

i.e. the intensity averaged with respect to ω and ν. Now the rate of absorption of photons per unit volume is also equal to $B_{12} n_1 \rho_\nu$ where ρ_ν is the energy density of radiation at frequency ν within the volume dv. Since $\rho_\nu = 4\pi \overline{I_\nu} c^{-1}$, we therefore have

$$4\pi Sn\overline{I_\nu}(h\nu)^{-1} = B_{12} n_1 4\pi c^{-1}\overline{I_\nu} \tag{5.23}$$

Referring again to fig. 5.7 equilibrium between the population of the levels requires

$$n_1(B_{12} 4\pi\overline{I_\nu} c^{-1} + b_{12}) = n_2(A_{21} + a_{21}) \tag{5.24}$$

Substituting in (5.24) for n_1 and n_2 from (5.21) and (5.23) results in an expression for the source function

$$J_\nu = \frac{\overline{I_\nu} + (cb_{12}A_{21}/4\pi a_{21}B_{12})\phi}{1 + \phi} \tag{5.25}$$

where $\quad \phi = a_{21}/A_{21}$

In the limit where local thermodynamic equilibrium applies collisional activation and deactivation dominates so $\phi \to \infty$ and $J_\nu \to B_\nu$, the Planck function. The quantity in brackets in (5.25) must therefore be B_ν and

$$J_\nu = \frac{\overline{I_\nu} + \phi B_\nu}{1 + \phi} \tag{5.26}$$

Note that as $\phi \to 0, J_\nu \to \overline{I_\nu}$ and we have isotropic scattering.

If ψ is written for the heating rate, or the rate of increase in energy per unit volume due to radiation processes

$$\psi = -\int_{\Delta\nu} \int_{4\pi} \frac{dI_\nu}{ds} \, d\omega \, d\nu \tag{5.27}$$

Substituting for dI/ds from the equation of transfer (2.3) (with J_ν written for B_ν):

$$\psi = \int_{\Delta\nu} \int_{4\pi} k_\nu n(I_\nu - J_\nu) \, d\omega \, d\nu$$

$$= 4\pi S n (\overline{I_\nu} - J_\nu) \tag{5.28}$$

from (5.22). Substituting for $\overline{I_\nu}$ from (5.26) we have

$$J_\nu - B_\nu = \frac{\psi}{4\pi S n \phi} \tag{5.29}$$

Equation (5.29) expresses the difference between the source function and the black-body function in terms of the heating rate and the ratio ϕ of the probability of relaxation from the upper state through collisional processes a_{21} to that through radiation processes. As we should expect from the discussion at the beginning of this section, $J_\nu \simeq B_\nu$, i.e. LTE exists when the heating rate is small or when collisional processes dominate, i.e. when ϕ is large.

The most important application in the atmosphere of the theory of non-LTE radiative transfer is to emission by the 15 μm vibration–rotation band of carbon dioxide, which for the LTE case was considered in §4.6. With the same approximation as was employed there, that cooling to space is the dominant term, for the non-LTE case (4.20) applies with B_ν replaced by J_ν i.e.

$$-\psi = \int_{\Delta\nu} \pi J_\nu \frac{d\tau_\nu^*(z, \infty)}{dz} \, d\nu \tag{5.30}$$

with

$$\tau_\nu^*(z, \infty) = \exp\left(-\int_z^\infty \frac{5}{3} k_\nu n \, dz'\right) \tag{5.31}$$

Over the spectral interval in which the band is contained, J_ν may be considered constant and

$$-\psi = \tfrac{5}{3} \pi J_\nu n S \overline{\tau_\nu^*} \tag{5.32}$$

where the quantity

$$\overline{\tau_\nu^*} = \frac{1}{S} \int_{\Delta\nu} k_\nu \tau_\nu^*(z, \infty) \, d\nu \tag{5.33}$$

is the probability of a photon emitted by a carbon dioxide molecule at level z getting out to space. Substituting for J_ν from (5.29) in (5.32) we have

$$-\psi = \frac{\tfrac{5}{3} \pi S n \overline{\tau_\nu^*} B_\nu}{1 + 5\overline{\tau_\nu^*}/12\phi} \tag{5.34}$$

Under conditions of LTE, the second term in the denominator of (5.34) is zero. The degree to which the cooling rate, therefore, differs from its LTE value depends not only on ϕ but also on $\overline{\tau_\nu^*}$ the transparency of the atmosphere to space.

For the 15 μm carbon dioxide band, and collisions with air molecules, the collisional relaxation time $a_{21}^{-1} \simeq 3 \times 10^{-5}$ s at standard pressure and 210 K (an

61

average temperature for the mesosphere). Since a_{21} is mainly dependent on the collision frequency it is approximately proportional to pressure. The radiative life time $A_{21}^{-1} = 0.74\,\mathrm{s}$. We therefore have $\phi \simeq 2.4 \times 10^4 p$ (where p is pressure in atm) and $\phi = 1$ at a pressure of 0.04 mb or a height of 73 km. Because, at this level $\overline{\tau_\nu^*}$ is small (problem 5.12), the LTE approximation applies to somewhat higher altitudes, in fact to about the 80 km level. Above 80 km the second term in the denominator becomes more and more significant. For instance at the 10^{-6} atm level (~ 96 km), the rate of cooling is only 0.05 of what it would be under LTE (problem 5.14 and fig. 5.9).

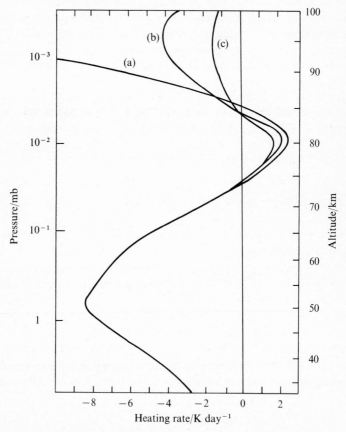

Fig. 5.9. Heating rates for the 15 μm band of carbon dioxide for a mean atmosphere, with collisional relaxation times at standard pressure (a) 2×10^{-6} s, (b) 1×10^{-5} s, (c) 3×10^{-5} s (after Williams, 1971).

In the above discussion, the cooling to space approximation has been used. We conclude this section by writing down the heating rate equations in a particularly useful form for numerical computation in which the radiative transfer between all layers is included correctly.

Consider the atmosphere divided into a number of discrete layers. With the plane parallel approximation of §4.3 and for a frequency range sufficiently wide to include much or all of a vibration–rotation band, but sufficiently narrow that B_ν or J_ν is constant within the range, (2.4) and (4.6) can be written in finite difference form

$$\psi_k = \sum_j C_{kj} J_{\nu j} + \psi_{Sk} \tag{5.35}$$

where ψ_k is the heating rate at level k, $J_{\nu j}$ is the source function at level j, C_{kj} is an element of a matrix known as the *Curtis matrix* which depends on atmospheric transmission and the summation is over all atmospheric layers. Heating due to solar absorption ψ_{Sk} has been included to make the treatment as general as possible. In the same notation (5.29) is

$$J_k = B_k + E_k \psi_k \tag{5.36}$$

where $E_k = (4\pi S n \phi)^{-1}$ appropriate to level k. In matrix notation, (5.35) and (5.36) are

$$\boldsymbol{\psi} = \mathbf{CJ} + \boldsymbol{\psi}_S$$

$$\mathbf{J} = \mathbf{B} + \mathbf{E}\boldsymbol{\psi} \tag{5.37}$$

which may be solved to give

$$\boldsymbol{\psi} = (\mathbf{I} - \mathbf{CE})^{-1}(\mathbf{CB} + \boldsymbol{\psi}_S) \tag{5.38}$$

$\mathbf{I}$ being the unit matrix.

In (5.38) $\mathbf{C}$ describes atmospheric transmission between different levels which for a uniformly mixed gas such as carbon dioxide may be computed once and for all for a variety of standard atmospheric conditions (the dependence of transmission on temperature is small). The vector $\mathbf{B}$ describes a particular atmospheric temperature profile. The form of (5.38), therefore, is particularly convenient for numerical computation of radiative heating rates for different atmospheric conditions.

Problems

5.1   Assuming different uniform temperatures above the turbopause at 120 km, estimate for different levels of solar activity the level above which the most abundant constituent is helium.

5.2   Show that a body of cross-sectional area A travelling at velocity V through a gaseous medium of density ρ experiences a drag force equal to $\frac{1}{2}\rho A c_D V^2$ where c_D is the drag coefficient. Hence estimate for standard atmospheric conditions the change in orbital period per orbit for a satellite of mass 100 kg with $A = 1$ m^2, $c_D = 2$, in a circular orbit at 600 km altitude. Observation of the details of satellite orbits is an important method of measuring density in the thermosphere.

5.3   Calculate the value of β (5.9) for helium atoms for $T = 1000$ K. Estimate the temperature which would be necessary for the value of β for helium to be 1.6 m s^{-1}, i.e. equal to that for hydrogen at $T = 1000$ K.

5.4   Show that, from simple kinetic theory, the diffusion coefficient D of hydrogen atoms present in small proportion in air at temperature T and total number density n is proportional to $T^{1/2}n^{-1}$.

Hence show that if a constant scale height H is assumed for the atmosphere as a whole above the turbopause where the total number density is n_0 then the density of hydrogen atoms is given by the differential equation

$$\frac{dn_H}{dz} + \frac{n_H}{H_H} = -\frac{\dot{N}n_0}{CT^{1/2}} \exp\left[\frac{-(z-z_0)}{H}\right] \tag{5.39}$$

where C is a constant.

5.5   Integrate (5.39). From your solution, given that n_H (120 km) = 2×10^{11} m^{-3}, $C = 2 \times 10^{20}$ m^{-1} s^{-1} K$^{-1/2}$ and finding suitable values for the other quantities from tables or diagrams, find the value of n_H at the escape level and hence the rate of escape of hydrogen atoms.

5.6   It has been suggested that oxygen present in the earth's atmosphere has resulted from the dissociation of water molecules in the high atmosphere followed by the escape of hydrogen. Assuming constant conditions calculate the loss of hydrogen during the earth's history ($\sim 4.5 \times 10^9$ years) and test the plausibility of this hypothesis. (This possible source is now believed to be small compared with oxygen produced through photosynthesis.)

5.7   Solve (5.11) numerically in the following way (you will need to write a simple computer programme). Assume a pure oxygen atmosphere above a lower boundary at $z = 120$ km, $T = 300$ K and calculate from the hydrostatic equation the number density of atomic oxygen as a function of altitude for an isothermal atmosphere. Assume an absorption cross-section for atomic oxygen of 1.2×10^{-21} m^2, a value of the product ϵI_∞ of 1.1×10^{-3} W m^{-2}, a value of $\lambda = AT^{1/2}$ where $A = 3.6 \times 10^{-3}$ J m^{-1} s^{-1} K$^{-3/2}$ and solve (5.11) for layers beginning with the lower boundary. From the resulting values of T, recalculate a new density profile and continue the iteration until a satisfactory solution has been obtained.

5.8   For the temperature profile of appendix 6 calculate the total potential energy (thermal and gravitational) above 120 km as given by (3.19). Compare this with the energy absorbed during an eight hour period (assuming the sun overhead). Hence make a crude estimate of the amplitude of diurnal variation in the exosphere. Compare your estimate with fig. 5.4.

5.9   Associated with the equilibrium of ozone as described in §5.5, there are two very different time constants, the first being associated with changes in the relative amount of O and O_3, the combined number densities of O and O_3 being kept constant, and the second being associated with changes in the sum of the amounts of O and O_3, the ratio between them during the change being given by (5.17). Write down differential equations for changes under these two approximations and hence deduce expressions for the two time constants. Calculate values for them at 40 km, 30 km and 20 km given that $k_2 = 10^{-33}\,cm^6\,s^{-1}$, $k_3 = 10^{-15}\,cm^3\,s^{-1}$. Take values of J_2, J_3 and n_3 from figs. 5.5 and 5.6.

Note that at the lower levels the second time constant is very long indeed so that the ozone mixing ratio becomes a very good 'tracer' of atmospheric motions.

5.10   Measurements of the OH airglow show that approximately 10^{12} photons s^{-1} of wavelength between 1 μm and 3 μm are emitted from a 1 cm^2 column of atmosphere at levels near the mesopause. Assuming that the emission is uniformly distributed over the region from 75 to 95 km altitude, estimate the cooling rate in K day^{-1} due to this emission.

5.11   If W is the equivalent width of an absorption band for the path between level z and space (absorber path length $u = |\int \frac{5}{3}c\rho\,dz|$), show that $\overline{\tau_\nu^*}$ in (5.33) is given by

$$\overline{\tau_\nu^*} = \frac{1}{S}\frac{dW}{du}$$

where S is the strength of the band.

Hence show that for a band of a gas present in constant mixing ratio consisting of non-overlapping collision broadened lines under either the 'weak' or 'strong' approximation the probability of a photon emitted from a level z getting out to space is independent of height.

5.12   Calculate the value of $\overline{\tau_\nu^*}$ for a photon from the 15 μm carbon dioxide band under the non-overlapping 'strong' collision broadened approximation given that $\Sigma\,(S_i\gamma_{0i})^{1/2} = 1600\,cm^{-1}(g\,cm^{-2})^{-1/2}$ and $S = \Sigma\,S_i = 1.26 \times 10^5\,cm^{-1}(g\,cm^{-2})^{-1}$ and that the mass mixing ratio of carbon dioxide = 4.7×10^{-4}.

5.13   Consider cooling from the 15 μm carbon dioxide band at the level where the pressure is 10^{-6} atm ($\sim$ 96 km). Here $\phi \ll 1$ and $\overline{\tau_\nu^*} \approx 1$. Substitute for B_ν and for $A_{21}(A_{21} = 8\pi\nu^2 g_1 S/c^2 g_2$ where g_1 and g_2 are the statistical weights of the lower and upper levels respectively) in (5.34). Hence show that

$$-\psi = nh\nu\left\{\frac{g_2 a_{21}}{g_1[\exp(h\nu/kT) - 1]}\right\}$$

Notice that this result is independent of the band strength S, and that the quantity multiplying $nh\nu$ is the probability per unit time of a quantum of energy being acquired on collision. Put in values appropriate to the 10^{-6} atm level and find the cooling rate in K per day.

5.14   Compare the value found in problem 5.13 with that which would occur if thermodynamic equilibrium prevailed.

5.15   Check the assumption $\tau_\nu^* = 1$ for the lines in the 15 μm carbon dioxide band considered in problem 5.13 by finding the transmission at the centre of a typical line of strength $s = 10^3\,cm^{-1}(g\,cm^{-2})^{-1}$ between the level 10^{-6} atm and space. (The absorption coefficient at the centre of a Doppler broadened line is $s\pi^{-1/2}\gamma_D^{-1}$ – see problem 4.8 for an expression for γ_D).

5.16   Calculate for carbon dioxide at STP (molecular diameter 4×10^{-10} m) the number of collisions per molecule per second from simple kinetic theory considerations. If the relaxation time for de-excitation of the ν_2 vibration by collision is 6×10^{-6} s at STP, what is the probability of de-excitation on one collision?

This very small probability is because $h\nu$ for the ν_2 vibration is considerably greater than the average kinetic energy of a molecule ($\sim 3/2\,kT$) at ordinary temperatures so that being able to transfer enough energy from translational motion to vibrational motion in any one collision is very unlikely.

6
Clouds

6.1 Cloud formation

It was demonstrated in §3.2 that ascent of damp air can lead to condensation and hence cloud formation. Four main kinds of clouds can be distinguished according to the type of ascending motions which produce them: (1) layer clouds formed by widespread, regular ascent such as occurs near the boundary between air masses having different characteristics (frontal zones) or because of orography; (2) layer clouds formed by vertical mixing (cf. problem 3.9); (3) convective clouds; (4) clouds formed under stable conditions by limited vertical motion as air passes over hills or mountains. The two first processes lead to *stratus* clouds or, when at higher levels, *cirrus* clouds, convective processes produce *cumulus* clouds, and associated with mountains are typically *lenticular* or *wave* clouds.

Suitable nuclei on which condensation can occur are reasonably abundant in the atmsophere. Nuclei on which the freezing of a droplet may commence are, however, less abundant; the tops of clouds extending above the freezing level commonly contain supercooled droplets at temperatures well below 0 °C.

The presence of clouds influences the energetics of the atmosphere in two main ways: (1) by the part which clouds play in the atmospheric water cycle; latent heat is released on condensation and liquid water is removed from the atmosphere on precipitation; (2) by scattering, absorption and emission of solar and terrestrial radiation clouds influence very strongly the atmosphere's radiation budget. In this chapter these two main influences will be discussed in turn.

6.2 The growth of cloud particles

A cloud particle (droplet or ice crystal) has a typical diameter of $\sim 10\,\mu$m whereas a rain drop of sufficient size to precipitate is ~ 1 mm in diameter. Each precipitating particle, therefore, contains the same amount of liquid water as about one million cloud particles. Two possible processes are available for particle growth: (1) by diffusion of water vapour to the cloud particles and subsequent condensation; (2) by collision and coalescence of particles.

First consider growth by condensation of an isolated spherical particle of

67

mass m, radius r, density ρ_L in an environment where at distance x from the particle the vapour density is ρ. The rate of particle growth is sufficiently slow for a steady state diffusion equation to be applied to it in which the mass of water vapour crossing any spherical surface in unit time is independent of x and equal to $\dot{m}$, the rate of increase of mass of the particle.

From Fick's law of diffusion we have, therefore,

$$\dot{m} = 4\pi x^2 D \frac{d\rho}{dx} \tag{6.1}$$

where D is the diffusion coefficient of water vapour in air. Integrating (6.1) from the surface of the drop where the vapour density is ρ_r to a large distance away where it is ρ_∞ we have

$$4\pi D \int_{\rho_\infty}^{\rho_r} d\rho = \int_\infty^r \frac{\dot{m}}{x^2} dx \tag{6.2}$$

As water vapour condenses on the particles, latent heat is released at a rate $L\dot{m}$ where L is the latent heat of condensation. This has to be conducted away. A temperature gradient dT/dx is therefore, set up, the equation describing the conduction being similar to (6.1), namely

$$L\dot{m} = -4\pi x^2 \lambda \frac{dT}{dx} \tag{6.3}$$

where λ is the thermal conductivity of air. Equation (6.3) may be integrated to give

$$L\dot{m} = 4\pi \lambda r (T_r - T_\infty) \tag{6.4}$$

where T_r and T_∞ are respectively the temperatures at the surface of the drop and at a large x.

From (6.2) and (6.4) the rate of droplet growth under different conditions can be evaluated (problems 6.1, 6.2 and 6.3). The process is particularly important at temperatures $\sim -10\,^\circ C$ when a few ice crystals are present in a cloud which predominantly contains supercooled water drops. The saturation vapour pressure over the water drops (see appendix 2) is about 10% more than that over the ice crystal which, therefore, grows at the expense of the water drops.

The other process of droplet growth is by collision and coalescence for which the rate of growth of a larger droplet of fall speed V in a population of smaller drops of fall speed v will be

$$\frac{dr}{dt} = \frac{\epsilon w}{4\rho_L} (V - v) \tag{6.5}$$

where w is the mass of liquid water in the form of smaller droplets per unit volume of the cloud and ϵ is the collision efficiency. Since small droplets which approach each other tend to follow the streamlines of the air, they are

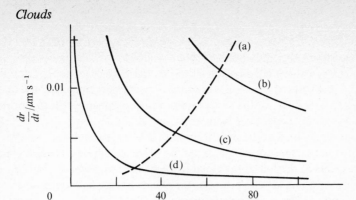

Fig. 6.1. Rate of growth of cloud droplets for cloud with liquid water content $1\,\mathrm{g\,m^{-3}}$ (a) by coalescence assuming that distribution function of cloud droplet radii r given by $4r^2r_m^{-3}\exp(-2r/r_m)$ with $r_m = 2\,\mu\mathrm{m}$, (b), (c), (d) by condensation assuming water vapour pressure is above saturation by (b) 0.04 mb, (c) 0.012 mb, (d) 0.0025 mb: after Shishkin (see Matveev, 1967, p. 513).

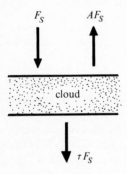

Fig. 6.2

unlikely to collide unless they are both large enough and unless there is sufficient size differential between them. Cloud particles grow to $30\,\mu\mathrm{m}$ in diameter largely by condensation processes after which the collision efficiency becomes larger, subsequent growth continuing by coalescence (fig. 6.1).

6.3 The radiative properties of clouds

In chapter 2 when considering the simple radiative equilibrium model the assumption was made that solar radiation incident on the atmosphere is unmodified by the atmosphere and all absorbed at the surface; although the absorption of a small proportion of the solar radiation by ozone in the ultraviolet and by water vapour in the infrared were mentioned (cf. fig. 2.1). Solar radiation is also modified on its passage through the atmosphere by scattering processes. At a wavelength of 0.5 μm about 10% of solar radiation is scattered by molecules in the atmosphere because of Rayleigh scattering (problems 6.14, 6.15 and 6.16). Clouds are also of great importance; because on average they cover $\sim 50\%$ of the earth's surface their influence on both the incoming and outgoing radiation is large.

Considering first the modification of solar radiation by clouds (fig. 6.2), if F_S is the incident solar flux on the top of the cloud, AF_S the reflected upward flux at the top of the cloud (A being the *cloud albedo*) and τF_S the downward flux at the bottom of the cloud (τ being its *transmissivity*), the amount of radiant energy absorbed is $F_S(1 - A - \tau)$. For a typical stratus cloud and for radiation integrated over the whole spectrum, typical values are $A = 0.5, \tau = 0.3$. Calculations of how A and τ depend on cloud thickness and on the properties of the cloud particles are presented in §6.4.

Because liquid water and ice absorb strongly throughout the infrared region, clouds are strong absorbers of terrestrial radiation. For considerations of the atmosphere's radiation budget it is satisfactory to assume that all clouds except cirrus absorb and emit terrestrial radiation as black bodies. Cirrus clouds are often thin and tenuous; although their albedo is small their transmissivity and emissivity may vary very considerably.

6.4 Radiative transfer in clouds

In chapters 2 and 4 radiative transfer theory was developed considering absorption and emission only. A simple extension of the theory enables scattering also to be taken into account.

Considering scattering particles distributed uniformly through a volume, a scattering coefficient σ can be defined in the same way as the absorption coefficient (2.1) so that the element of optical depth $d\chi = (k + \sigma)\rho\,dz$.

The absorption coefficient k takes into account absorption both by the particles and by the gas in between them. The coefficient σ accounts for scattering in all directions. A quantity called the *albedo for single scattering* ω is the ratio $\sigma/(k + \sigma)$.

As in chapters 2 and 4 we are considering transfer in the vertical dimension only (integrating over other angular dependences) with fluxes $F^\uparrow$ and $F^\downarrow$. The transfer equations become

$$\frac{dF^\downarrow}{d\chi^*} + F^\downarrow = \pi B(1 - \omega) + \omega\{fF^\downarrow + (1 - f)F^\uparrow\} \tag{6.6}$$

$$-\frac{dF^\uparrow}{d\chi^*} + F^\uparrow = \pi B(1 - \omega) + \omega\{(1 - f)F^\downarrow + fF^\uparrow\} \tag{6.7}$$

The first term on the r.h.s. is the emission term depending on the black-body function B in exactly the same way as in (2.3). The second term is the contribution to radiation transfer from the scattering processes. The fraction f is the proportion of the scattered radiation which in one scattering event goes into a forward direction while $(1 - f)$ is that which is scattered in a backward direction. For isotropic scattering $f = \frac{1}{2}$.

Equations (6.6) and (6.7) account properly for absorption, emission and for multiple scattering in cases where the restriction to one dimension is a satisfactory approximation. Their solution in the general case can be complicated. Here we shall treat only the case for solar radiation incident vertically downwards on the top of a cloud; the emission term (the first on the r.h.s.) in (6.6) and (6.7) will therefore be omitted.

It is instructive first to let $\omega = 1$ when we have pure scattering. Equations (6.6) and (6.7) may then be written

$$\left.\begin{aligned}
\frac{dF^{\downarrow}}{d\chi^*} + (F^{\downarrow} - F^{\uparrow})(1 - f) = 0 \\
\frac{dF^{\uparrow}}{d\chi^*} + (F^{\downarrow} - F^{\uparrow})(1 - f) = 0
\end{aligned}\right\} \tag{6.8}$$

giving immediately

$$F^{\downarrow} - F^{\uparrow} = \text{constant}$$

and for a uniform cloud of optical thickness χ_0^* the boundary conditions are (fig. 6.2) $F^{\downarrow}(\chi^* = 0) = F_S$, $F^{\uparrow}(\chi^* = \chi_0^*) = 0$ so that the cloud albedo A and transmissivity τ respectively become

$$A = \frac{\chi_0^*(1 - f)}{1 + \chi_0^*(1 - f)} \tag{6.9}$$

$$\tau = \frac{1}{1 + \chi_0^*(1 - f)} \tag{6.10}$$

Note that $A + \tau = 1$ as is required by conservation of energy.

When $\omega \neq 1$ (6.6) and (6.7) may be solved by differentiating again (6.6) and then substituting for $dF^{\uparrow}/d\chi^*$ from (6.7). The solution is

$$F^{\downarrow} = C \exp(-\alpha\chi^*) + D \exp[-\alpha(\chi_0^* - \chi^*)]$$

with a similar expression for $F^{\uparrow}$ where

$$\alpha^2 = (1 - \omega)(1 + \omega - 2\omega f) \tag{6.11}$$

and C and D are constants. Applying the boundary conditions leads to

$$A = \beta \frac{1 - \exp(-2\alpha\chi_0^*)}{1 - \beta^2 \exp(-2\alpha\chi_0^*)} \tag{6.12}$$

$$\tau = \frac{(1 - \beta^2) \exp(-\alpha\chi_0^*)}{1 - \beta^2 \exp(-2\alpha\chi_0^*)} \tag{6.13}$$

Clouds

where
$$\beta = \frac{\alpha - 1 + \omega}{\alpha + 1 - \omega} \tag{6.14}$$

Note that as χ_0^* becomes large

$$A \to \beta \quad \text{and} \quad \tau \to 0$$

The quantity β, therefore, is the albedo of a very thick cloud. For $\omega = 0.9997$, $f = 0.9$, values appropriate at a wavelength of $\sim 0.7\,\mu m$ to a stratus cloud with drops $\sim 10\,\mu m$ diameter, the value of β is 0.925. The optical thickness of the cloud χ_0^* needs to be large for its albedo to approach this value of β; for $\chi_0^* \simeq 125$ for instance, the albedo of this cloud $A = 0.90$. For a typical stratus cloud 0.5 km thick, $\chi_0^* \simeq 20$, giving $A \simeq 0.66$ (problem 6.9).

Problems

(See the appendices for some of the physical data necessary for the solution of problems.)

6.1   Integrate the Clausius Clapeyron equation (3.13) for water vapour between temperatures T_r and $T_\infty (T_r - T_\infty \ll T_\infty)$ to give an expression in the following form for the corresponding saturation vapour pressures $p_s(T_r)$ and $p_s(T_\infty)$

$$\ln\left[\frac{p_s(T_r)}{p_s(T_\infty)}\right] \simeq \frac{LM_r(T_r - T_\infty)}{RT_\infty^2} \tag{6.15}$$

(R is the gas constant and M_r molecular weight of water.)

6.2   Write (6.2) in terms of vapour pressure rather than vapour density. Using (6.4) and (6.15) derive the following expression for the rate of growth of a water droplet.

$$r\frac{dr}{dt} \simeq \frac{S - 1}{[(L^2 M_r \rho_L / \lambda R T^2) + \{\rho_L R T / p_s(T_\infty) D M_r\}]} \tag{6.16}$$

where $S = p_\infty / p_s(T_\infty)$, p_∞ being the actual vapour pressure far away from the droplet. $S - 1$ ($\ll 1$) is the *supersaturation* of the vapour phase.

The effective degree of supersaturation depends also on (1) the droplet radius especially for drops $< 1\,\mu m$ in radius over which the equilibrium vapour pressure is considerably higher than over a plane surface, and (2) the purity of the water, the equilibrium vapour pressure being less for salt solutions. Because of these considerations very small drops only persist on hygroscopic nuclei.

6.3   For pure water use (6.16) to calculate the time taken for a droplet to grow from $2\,\mu m$ radius (1) to $10\,\mu m$ radius and (2) to $40\,\mu m$ radius by condensation if the supersaturation is 0.05% (i.e. $S - 1 = 5 \times 10^{-4}$) and $T = 273$ K. ($D = 0.23$ cm^2 s^{-1}.)

6.4   From Stokes' law for the viscous force F acting on a sphere of radius r falling with terminal velocity v in a fluid of viscosity η namely

72

$$F = 6\pi\eta v r$$

calculate the time taken for drops of radius $1\,\mu m$, $10\,\mu m$, $100\,\mu m$, to fall through 1 km. (η for air $= 1.7 \times 10^{-5}\,\mathrm{kg\,m^{-1}\,s^{-1}}$.)

6.5 From values of the saturation vapour pressure over plane liquid water surfaces and plane ice surfaces given in appendix 2 for $-10°C$, calculate the time taken for an ice crystal in a cloud of water droplets at $-10°C$ to grow from $1\,\mu m$ radius to $100\,\mu m$ radius, assuming that the crystal remains spherical.

Calculate also the temperature of the ice crystal. (Latent heat of sublimation of ice $= 2800\,\mathrm{J\,g^{-1}}$, $D = 0.23\,\mathrm{cm^2\,s^{-1}}$.)

6.6 For $\omega = 1$, find A and τ for $\chi_0^* = 20$ and $f = 0.9$. Compare with the values quoted in the chapter when $\omega = 0.9997$.

6.7 From (6.11) and (6.14) show that

$$\beta^2 = \frac{1 - \omega f - \alpha}{1 - \omega f + \alpha}$$

6.8 Derive (6.12) and (6.13)

6.9 For stratus cloud described in text with $\chi_0^* = 20$, find transmissivity τ. Also find the absorptivity of the cloud (i.e. $1 - \tau - A$).

6.10 The scattering cross-section at wavelength λ of a drop of radius r when the ratio $2\pi r/\lambda \gg 1$ is approximately $2\pi r^2$. Find the optical thickness of a cloud 0.5 km thick containing $150\,\mathrm{drops\,cm^{-3}}$ having $r = 5\,\mu m$. Remember $\chi_0^* \hat{=} 1.66\chi_0$.

6.11 At $2.3\,\mu m$, for drops having $r = 7\,\mu m$, $\omega = 0.988$, $f = 0.9$, find β and find the value of χ_0^* to give an albedo of 0.98β. Compare this value of χ_0^* with that for typical stratus.

6.12 At $10.6\,\mu m$, the wavelength of the CO_2 laser, $\omega = 0.36$, $f = 0.9$. Find β and the transmissivity of diffuse radiation at this wavelength in 100 m of fog having the same properties as the stratus cloud in the text. (Assume χ_0^* is the same at $10.6\,\mu m$ as at $0.7\,\mu m$.)

6.13 In the visible part of the spectrum the albedo of the Venus clouds is approximately 0.9. Supposing the clouds to be very deep and that scattering is isotropic, what is the minimum value of ω for the cloud particles?

6.14 The scattering coefficient σ_R for Rayleigh scattering at wavelength λ by a gas containing N molecules per unit volume is

$$\sigma_R = \frac{32\pi^3}{3N\lambda^4} (n - 1)^2$$

where n is the refractive index. From the data in appendix 1, calculate the extinction by Rayleigh scattering for a column of atmosphere at wavelengths of 0.3, 0.6 and $1\,\mu m$.

6.15 Explain why scattered sunlight observed at $90°$ to the direction of the sun shows very strong polarization. Why is the polarization only about 90%, not 100%?

6.16 Why does the sky appear blue while clouds in general appear white?

7
Dynamics

7.1 Total and partial derivatives

So far in making simple atmospheric models we have considered energy transfer by radiation, conduction and convection. In chapter 3 the atmosphere's radiative sources and sinks were shown to be driving a thermodynamic engine which generates kinetic energy as well as potential energy. Thermodynamic considerations alone cannot tell us much more about the form of this kinetic energy. In this chapter and those that follow, therefore, we turn our attention to the atmosphere's motions, our aim being to find simple descriptions or physical models which can aid our understanding of some of the main features of the atmosphere's circulation.

When applying the laws of motion to an element of fluid, it is important to realize that the fluid being considered is in general moving with respect to our chosen frame of reference. We shall frequently need to relate the derivative with respect to time t of a scalar quantity ψ taken following the fluid, i.e. the *individual* or *total derivative* $d\psi/dt$ to the *local*, or *partial*, *derivatives* $\partial\psi/\partial t$ etc. appropriate to a fixed point in the chosen frame of reference.

Since ψ is varying in both space and time

$$d\psi = \frac{\partial\psi}{\partial t} dt + \frac{\partial\psi}{\partial x} dx + \frac{\partial\psi}{\partial y} dy + \frac{\partial\psi}{\partial z} dz$$

i.e.
$$\frac{d\psi}{dt} = \frac{\partial\psi}{\partial t} + u\frac{\partial\psi}{\partial x} + v\frac{\partial\psi}{\partial y} + w\frac{\partial\psi}{\partial z}$$

(7.1)

or
$$\frac{d\psi}{dt} = \frac{\partial\psi}{\partial t} + \mathbf{V} \cdot \text{grad } \psi$$

where
$$u = \frac{\partial x}{\partial t}, \quad v = \frac{\partial y}{\partial t}, \quad w = \frac{\partial z}{\partial t}$$

are the components of the fluid's velocity $\mathbf{V}$ with respect to Cartesian axes in our frame of reference.

Equation (7.1) will be used constantly in the paragraphs that follow.

7.2 Equations of motion

Newton's second law of motion applied to an element of fluid of density ρ

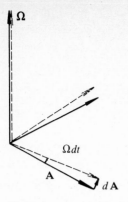

Ωdt

A

d**A**

Fig. 7.1

moving with velocity **V** in the presence of a pressure gradient ∇p and a gravitational field (described by **g′**) is (for derivation and discussion of this equation see for instance Batchelor (1967) or Rutherford (1959))

$$\frac{d\mathbf{V}}{dt} = \mathbf{g}' - \frac{1}{\rho}\nabla p + \mathbf{F} \tag{7.2}$$

where **F** is the frictional force on the element which will be discussed more fully in chapter 9.

Equation (7.2) applies to an absolute or inertial frame of reference (i.e. say fixed with respect to the solar system). Our interest is in motion relative to axes fixed with respect to the earth's surface, which is rotating with angular velocity Ω.

A vector **A** in frame Σ which is rotating at angular velocity Ω with respect to frame Σ' will have a component of motion $\Omega \wedge \mathbf{A}$ in frame Σ' due to the relative motion of the two frames (fig. 7.1) so that

$$\left(\frac{d\mathbf{A}}{dt}\right)_{\Sigma'} = \left(\frac{d\mathbf{A}}{dt}\right)_{\Sigma} + \Omega \wedge \mathbf{A} \tag{7.3}$$

If we denote differentiation in the Σ' frame and quantities referred to that frame by primes, unprimed quantities being referred to the Σ frame, and let $\mathbf{A} = \mathbf{r}$ (the radius vector) and $d'\mathbf{r}/dt$ in turn we have

$$\mathbf{V}' = \mathbf{V} + \Omega \wedge \mathbf{r}$$

and

$$\frac{d'\mathbf{V}'}{dt} = \frac{d\mathbf{V}'}{dt} + \Omega \wedge \mathbf{V}'$$

$$= \frac{d\mathbf{V}}{dt} + 2\,\Omega \wedge \mathbf{V} + \Omega \wedge (\Omega \wedge \mathbf{r})$$

so that in the rotating frame (7.2) becomes

75

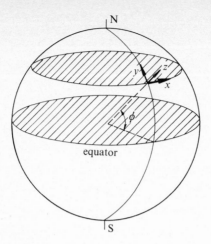

Fig. 7.2

$$\frac{d\mathbf{V}}{dt} + 2\boldsymbol{\Omega} \wedge \mathbf{V} + \boldsymbol{\Omega} \wedge (\boldsymbol{\Omega} \wedge \mathbf{r}) = -\frac{1}{\rho}\boldsymbol{\nabla}p + \mathbf{g}' + \mathbf{F}$$

or $$\frac{d\mathbf{V}}{dt} = 2\mathbf{V} \wedge \boldsymbol{\Omega} - \frac{1}{\rho}\boldsymbol{\nabla}p + \mathbf{g} + \mathbf{F} \qquad (7.4)$$

where $$\mathbf{g} = \mathbf{g}' - \boldsymbol{\Omega} \wedge (\boldsymbol{\Omega} \wedge \mathbf{r}) \qquad (7.5)$$

is the acceleration due to gravity and includes as it should the centrifugal term $\boldsymbol{\Omega} \wedge (\boldsymbol{\Omega} \wedge \mathbf{r})$ (problem 7.1).

The first term on the r.h.s of (7.4) is the Coriolis term which applies particularly to moving particles in rotating frames. It is perpendicular both to the direction of motion and the earth's axis of rotation. As we shall see in this and the following chapters, the Coriolis term has a profound influence on atmospheric motion.

A convenient set of axes at any point on the earth's surface (fig. 7.2) has x directed towards the east, y towards the north and z vertically upwards (or more precisely in the direction of the vector defined by (7.5). This set is not strictly Cartesian because the directions of the axes are functions of position on the spherical earth. If u, v, w are the components of the velocity $\mathbf{V}$ in the x, y, z directions respectively and $\mathbf{i}, \mathbf{j}, \mathbf{k}$ are unit vectors directed along each axis then taking into account the way the axes change with position (problem 7.2)

$$\frac{d\mathbf{V}}{dt} = \left(\frac{du}{dt} - \frac{uv\tan\phi}{a} + \frac{uw}{a}\right)\mathbf{i}$$

$$+ \left(\frac{dv}{dt} + \frac{u^2\tan\phi}{a} + \frac{wv}{a}\right)\mathbf{j} + \left(\frac{dw}{dt} - \frac{u^2 + v^2}{a}\right)\mathbf{k} \qquad (7.6)$$

where ϕ is the latitude and a the earth's radius. Also

$$
\begin{aligned}
2\mathbf{V} \wedge \boldsymbol{\Omega} = \; & 2\Omega(v \sin \phi - w \cos \phi)\mathbf{i} \\
& - 2\Omega u \sin \phi \; \mathbf{j} \\
& + 2\Omega u \cos \phi \; \mathbf{k}
\end{aligned}
\tag{7.7}
$$

The magnitudes of the various terms in (7.6) and (7.7) will be very different depending on the scale of the motion under study. In this chapter we shall be concerned with motions on what is generally known as the synoptic scale or larger, that is systems of typically 1000 km in horizontal dimension, very much larger than their vertical scale (of order 1 scale height or ~ 10 km). For this scale observed vertical velocities (typically 1 cm s^{-1}) are very much smaller than horizontal velocities (typically 10 m s^{-1}) so that, in the momentum equation (7.4) terms involving w can, to a first approximation, be neglected. Such motion is described as *quasi-horizontal*. Because of this, and because those terms in (7.6) which involve a in the denominator are smaller by about one order of magnitude than the other terms (problem 7.3) and may therefore again to a first approximation be neglected, the equation of motion (7.4) may be simplified to become

$$
\frac{d\mathbf{V}}{dt} = f\mathbf{V} \wedge \mathbf{k} - \frac{1}{\rho}\nabla p + \mathbf{F}
\tag{7.8}
$$

where $\quad f = 2\Omega \sin \phi$ $\qquad\qquad\qquad\qquad\qquad\qquad$ (7.9)

and where now

$$
\frac{d\mathbf{V}}{dt} = \mathbf{i}\frac{du}{dt} + \mathbf{j}\frac{dv}{dt}
$$

and all the terms are vectors in the horizontal plane.

For the vertical direction, under the same approximation (problem 7.3) the hydrostatic equation (1.4) applies.

We now consider two important approximations to (7.8) in which for different situations some of the terms may be neglected in comparison to the others. Other approximations are mentioned in the problems at the end of the chapter.

7.3 The geostrophic approximation

For large scale motion away from the surface the friction $\mathbf{F}$ is small. Further, for steady flow with small curvature $dV/dt \simeq 0$. The resulting motion is known as *geostrophic*. The geostrophic velocity $\mathbf{V}_g$ is given by (fig. 7.3)

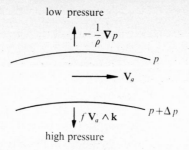

low pressure

high pressure

Fig. 7.3. The geostrophic approximation.

$$f\mathbf{V}_g \wedge \mathbf{k} = \frac{1}{\rho}\nabla p \qquad (7.10)$$

describing the familiar situation in which the wind blows parallel to the iso-bars, for the northern hemisphere in a clockwise direction around centres of high pressure (*anticyclones*) and anticlockwise around centres of low pressure (*depressions or cyclones*). For $\phi = 30°$, $\sin \phi = \frac{1}{2}$, $f = 7.29 \times 10^{-5}$ and for a pressure gradient of 2 mb $(100\,\text{km})^{-1}$, the appropriate velocity for balance is $23.8\,\text{m}\,\text{s}^{-1}$.

The geostrophic approximation works well at heights above ~ 1 km (at lower levels friction becomes important) and for latitudes $> \sim 10°$, so that under these conditions the wind velocity and direction may be deduced from a chart of isobars. Note that for horizontal motion, the Coriolis term is zero at the equator.

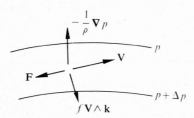

Fig. 7.4

At lower levels where friction is not negligible (cf. chapter 9), since friction acts in a direction approximately opposite to **V**, the velocity can no longer be precisely parallel to the isobars (fig. 7.4). For a high pressure centre the balancing velocity has a component outwards from the centre and for a low pressure centre an inward component. To conserve the mass of air within the system, air therefore sinks over high pressure systems and rises over low pressure ones (§9.5), producing an important modification to the vertical velocities which would occur in the absence of friction.

7.4 Cyclostrophic motion

At low latitudes f is small and for motion having a large curvature as for instance a tropical cyclone, another approximation to (7.8) is the *cyclostrophic* one in which the acceleration of the air towards the centre (distance r away) is balanced by the pressure gradient $\partial p/\partial n$

i.e.
$$\frac{V^2}{r} = \frac{1}{\rho}\frac{\partial p}{\partial n} \tag{7.11}$$

For a pressure gradient of $30\,\text{mb}\ (100\,\text{km})^{-1}$ and $r = 100\,\text{km}$, $V \simeq 50\,\text{m s}^{-1}$ – typical values for a tropical cyclone. Note that the sense of the motion is always cyclonic. If both the acceleration term and the Coriolis terms are included, the resulting solution is known as the *gradient wind* (problem 7.9).

The ratio of the acceleration term dV/dt to the Coriolis term is a dimensionless number known as the *Rossby number Ro*. For motion with speed V on a scale of typical horizontal dimension L,

$$Ro \simeq \frac{V^2/L}{fV} = \frac{V}{fL} \tag{7.12}$$

A typical value of Ro is 0.1 (problem 7.7) in which case the error introduced by the geostrophic approximation is $\sim 10\%$ (problem 7.8). The smallness of the Rossby number therefore is a measure of the validity of the geostrophic approximation.

7.5 Surfaces of constant pressure

If z is the height of a surface of constant pressure p on which two neighbouring points A and B have the same y co-ordinate

$$\frac{\partial p}{\partial x}\Delta x + \frac{\partial p}{\partial z}\Delta z = \Delta p = 0$$

Since
$$\frac{\partial p}{\partial z} = -g\rho \tag{1.2}$$

we have $g\dfrac{\partial z}{\partial x} = \dfrac{1}{\rho}\dfrac{\partial p}{\partial x}$

and similarly for

$\partial z/\partial y$

so that
$$g\rho\nabla_p z = \nabla_z p \tag{7.13}$$

where $\nabla_p z$ denotes the gradient of z in the surface of constant pressure. Equation (7.13) relates the gradient of pressure in a horizontal surface to the gradient of height in a nearby surface of constant pressure.

Dynamics

Because of the variation of g with altitude and with latitude (problem 7.11), (7.13) is not easy to use as it stands. It is convenient to define geopotential Φ as

$$\Phi = \int_0^z g \, dz$$

If now the geostrophic approximation (7.10) is written in terms of Φ, we have

$$\mathbf{V}_g = \frac{1}{f} \mathbf{k} \wedge \nabla_p \Phi \qquad (7.14)$$

Notice that the density ρ has disappeared from the equation; it is therefore applicable without change to any level in the atmosphere. For this reason standard meteorological charts are plotted as contours of Φ at various pressure levels rather than as contours of pressure on horizontal surfaces. It is usual to express Φ in terms of *geopotential height* Φ/g_0, where g_0 is the standard value of g at the surface (problem 7.11).

7.6 The thermal wind equation

Horizontal pressure gradients arise in the atmosphere owing to density differences which in turn are related to temperature gradients. To relate the geostrophic wind field and its variation with height to the temperature field (7.14) can be applied to two surfaces of constant pressure p_1 and p_2 so that

$$\mathbf{V}_g(p_2) - \mathbf{V}_g(p_1) = \frac{1}{f} \mathbf{k} \wedge \nabla_p (\Phi_2 - \Phi_1) \qquad (7.15)$$

Now, from the hydrostatic equation

$$\Phi_2 - \Phi_1 = \frac{R\bar{T}}{M_r} \ln \frac{p_1}{p_2} \qquad (7.16)$$

where $\bar{T}$ is the mean temperature between the surfaces so that

$$\mathbf{V}_g(p_2) - \mathbf{V}_g(p_1) = \mathbf{V}_t = \frac{R}{M_r f} \ln \frac{p_1}{p_2} \mathbf{k} \wedge \nabla_p \bar{T} \qquad (7.17)$$

The quantity $\mathbf{V}_t$ is called the *thermal wind* over the interval (p_1, p_2); it is related to the $\bar{T}$ field in a similar way to which $\mathbf{V}_g$ is related to the Φ field (fig. 7.5). The quantity $\Phi_2 - \Phi_1$, is known as the *thickness*; thickness charts are effectively charts of the mean temperature between different pressure levels.

Various illustrations of the use of the thermal wind are given in the problems at the end of the chapter. Fig. 5.1 shows the mean fields of zonal wind and temperature illustrating, for the atmosphere as a whole, how the two are related.

80

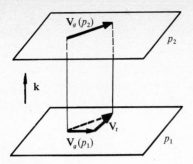

Fig. 7.5

7.7 The equation of continuity

A further equation we shall require in later chapters is the equation of continuity which states that the net flow of mass into unit volume per unit time is equal to the local rate of change of density.

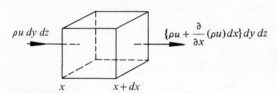

Fig. 7.6

In the elementary volume (fig. 7.6) where the density is ρ and u, v, w the velocity components along Cartesian axes, the mass entering the volume per unit time at x over the face of area $dy\,dz$ is $\rho u\,dy\,dz$, and leaving at $x + dx$ is $(\rho u + (\partial/\partial x)(\rho u)\,dx)\,dy\,dz$. Adding all the components together, we can write,

$$\frac{\partial(\rho u)}{\partial x} + \frac{\partial(\rho v)}{\partial y} + \frac{\partial(\rho w)}{\partial z} = -\frac{\partial\rho}{\partial t}$$

or $\quad\quad \text{div}\,\rho\mathbf{V} = -\frac{\partial\rho}{\partial t}$ $\hfill$ (7.18)

Equation (7.18) is the *equation of continuity*. For an incompressible fluid it reduces to

$$\text{div}\,\mathbf{V} = 0 \hfill (7.19)$$

A useful form of the continuity equation in which the pressure p is the vertical co-ordinate is found by considering an elemental column of atmosphere (fig. 7.7) confined between the surfaces of constant pressure p and $p - \delta p$. The mass of the column δm is equal to $\rho\,\delta x\,\delta y\,\delta z$ which by the hydrostatic equation gives

81

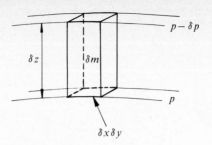

Fig. 7.7

$$\delta m = \frac{\delta x \, \delta y \, \delta p}{g}$$

Following the motion, the mass δm is conserved i.e.

$$0 = \frac{1}{\delta m}\frac{d}{dt}(\delta m) = \frac{g}{\delta x \, \delta y \, \delta p}\frac{d}{dt}\left(\frac{\delta x \, \delta y \, \delta p}{g}\right)$$

Carrying out the differentiation and taking the limits we find

$$\frac{\partial u}{\partial x} + \frac{\partial v}{\partial y} + \frac{\partial \omega}{\partial p} = 0 \qquad (7.20)$$

where $\omega = dp/dt$.

Equation (7.20) is the continuity equation in isobaric co-ordinates; note that, as with many other equations written in the isobaric system, it does not involve the density.

Problems

7.1 Calculate the value of the centrifugal acceleration of a particle at the equator and compare with g. What is the deviation of the plumb-line direction from the true direction of the centre of a uniform spherical earth at latitude 45°?

7.2 Show that

$$\frac{d\mathbf{V}}{dt} = \mathbf{i}\frac{du}{dt} + \mathbf{j}\frac{dv}{dt} + \mathbf{k}\frac{dw}{dt}$$

$$+ u\frac{d\mathbf{i}}{dt} + v\frac{d\mathbf{j}}{dt} + w\frac{d\mathbf{k}}{dt} \qquad (7.21)$$

Show that for the set of axes defined by fig. 7.2

$$\frac{d\mathbf{i}}{dt} = u\frac{\partial \mathbf{i}}{\partial x} = \frac{u}{a \cos \phi}(\mathbf{j} \sin \phi - \mathbf{k} \cos \phi) \qquad (7.22)$$

$$\frac{d\mathbf{j}}{dt} = u\frac{\partial \mathbf{j}}{\partial x} + v\frac{\partial \mathbf{j}}{\partial y} = -\frac{u\tan\phi}{a}\mathbf{i} - \frac{v}{a}\mathbf{k} \qquad (7.23)$$

$$\frac{d\mathbf{k}}{dt} = u\frac{\partial \mathbf{k}}{\partial x} + v\frac{\partial \mathbf{k}}{\partial y} = \frac{u}{a}\mathbf{i} + \frac{v}{a}\mathbf{j} \qquad (7.24)$$

Hence derive (7.6).

7.3 For synoptic scale motions of horizontal dimension $\sim 10^3$ km, of vertical dimension ~ 10 km for which a typical horizontal velocity is $10\,\text{m s}^{-1}$, a typical vertical velocity is $1\,\text{cm s}^{-1}$, estimate the relative magnitudes of the various terms in (7.6) and (7.7). Compare the magnitude of the terms in the vertical direction with g.

7.4 What is the pressure gradient required at the earth's surface at 45° latitude to maintain a geostrophic wind velocity of $30\,\text{m s}^{-1}$?

7.5 For motion around a centre of pressure 100 km away, at 30° latitude, compute the wind speed for which the Coriolis term will be equal to the acceleration term.

7.6 In the absence of a pressure gradient show that the radius of curvature of the flow is $-V/f$ and is anticyclonic. What is the period of a complete oscillation? Such flow is known as *inertial* flow and has been observed in the oceans as well as in the atmosphere.

7.7 Compute the value of the Rossby number for a typical case (latitude 45°, $L \simeq 1000$ km, $V \simeq 10\,\text{m s}^{-1}$). For the same case, what is the value of the Rossby number on Mars and Venus?

7.8 When $Ro = 0.1$ what is the error in the geostrophic wind approximation.

7.9 When friction is neglected in (7.8) show that when the other terms are retained, in the presence of a pressure gradient $\partial p/\partial n$ for flow with radius of curvature r (r is considered positive when the centre of curvature is in the positive n direction, i.e. cyclonic curvature is positive):

$$V = -\frac{fr}{2} \pm \left[\frac{f^2 r^2}{4} - \frac{r}{\rho}\frac{\partial p}{\partial n}\right]^{1/2} \qquad (7.25)$$

This approximation is known as the *gradient wind*.

Show that for anticyclonic curvature

$$\frac{\partial p}{\partial n} < \frac{\rho r f^2}{4} \qquad (7.26)$$

and hence that for anticyclones the pressure gradient decreases towards the centre. This is why the pressure gradients are small and the winds light near the centre of an anticyclone.

7.10 The atmospheric surface pressure at radius r_0 from the centre of a tornado rotating with constant angular velocity ω is p_0. Show that the surface pressure at the centre of the tornado is

$$p_0 \exp\left(-\omega^2 r_0^2/2RT\right)$$

when the temperature T is assumed constant.

Dynamics

7.11   The geopotential height z_g is defined such that

$$z_g = \frac{\Phi}{g_0} = \frac{1}{g_0} \int_0^z g \, dz \qquad (7.27)$$

where Φ is the geopotential and where g_0 has a standard value of 9.807 m s^{-2}, its mean value at the surface.

Derive an expression for the variation with height of the acceleration due to gravity. Over a place where the value of g at the surface ($z = 0$) is equal to g_0, what is the difference between z_g and the geometric height z when $z = 100$ km?

The value of g decreases by approximately 0.5% between the equator and the poles. Again for $z = 10$ km and 100 km, compute the differences in geopotential height z_g and geometric height z, over the equator and over the poles (cf. appendix 4).

7.12   On the 700 mb surface at latitude 45° the contours for 3300 m and 3500 m geopotential height are 1000 km apart. What is the magnitude of the geostrophic wind?

7.13   Fig. 7.8(a) is a cross-section of part of a front. At a certain level, the air possesses temperatures T_2 and T_1 respectively ($T_2 > T_1$) on the two sides of the front which has a slope of α relative to the horizontal. Apply the hydrostatic equation and the geostrophic wind equation to the region AB, together with the condition that the pressure must be continuous across the frontal surface. Show that in equilibrium the components of velocity v_2 and v_1 in the y direction on the two sides of the front satisfy the relation

$$(T_1 - T_2)g \tan \alpha = (v_2 T_1 - v_1 T_2)f \qquad (7.28)$$

If $T_2 - T_1 = 3$ K, $v_1 - v_2 = 10 \text{ m s}^{-1}$, find α. The slopes of typical fronts vary from around 1/50 to 1/300. Note that if $f = 0$, i.e. the earth were not rotating, the sloping surface could not be in equilibrium.

7.14   From the sense of the velocity change from (7.28), show that the kink in the isobars is always cyclonic (fig. 7.8(b)).

7.15   The wind at the surface is from the west. At cloud level it is from the south. Do you expect the temperature to rise or fall?

7.16   If the pole is 40 K colder than the equator and the surface wind is zero what wind would you expect at the 200 mb pressure level? Compare the temperature and wind fields of fig. 5.1 and show so far as you can that they satisfy (7.17).

7.17   At all levels from 1000 mb to 300 mb for the front shown in fig. 7.8(a) assume a temperature difference $T_2 - T_1$ of 5 K. For 45° latitude apply the thermal wind equation to the region between two places $x = 0$ and $x = 500$ km, the front being at the surface at $x = 0$ and at the 300 mb level at $x = 500$ km. If the surface wind is zero what wind speed will occur at the 300 mb level? This rather crude calculation shows that the presence of high upper level winds in the vicinity of a front roughly parallel to the frontal surface is consistent

84

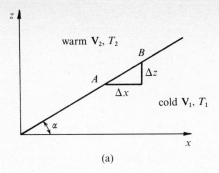

(a)

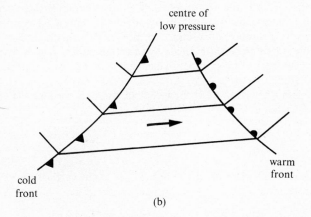

centre of
low pressure

warm
front

cold
front

(b)

Fig. 7.8. (a) Cross-section through a front. (b) Warm sector of typical depression showing warm and cold fronts.

with the thermal wind equation. The concentrated region of high winds in the upper troposphere associated with fronts is known as the *jet stream*.

7.18   For a horizontal temperature gradient in the y direction only, derive a differential form of the thermal wind equation (7.17) giving the shear with height of the wind component along the x axis, i.e.

$$\frac{\partial u}{\partial z} = -\frac{g}{Tf}\frac{\partial T}{\partial y} \qquad (7.29)$$

7.19   For horizontal flow (i.e. $w = 0$) show that

$$\text{div}_h\mathbf{V} = \boldsymbol{\nabla}_h\boldsymbol{\cdot}\mathbf{V} = \left(\mathbf{i}\frac{\partial}{\partial x} + \mathbf{j}\frac{\partial}{\partial y}\right)\boldsymbol{\cdot}(\mathbf{i}u + \mathbf{j}v)$$

Hence using some of the results of problem 7.2 show that

$$\text{div}_h\mathbf{V} = \frac{\partial u}{\partial x} + \frac{\partial v}{\partial y} - \frac{v}{a}\tan\phi \qquad (7.30)$$

7.20 From (7.30) show that if ρ is considered constant the divergence of the geostrophic wind $\mathbf{V}_g$ is

$$\text{div}_h \mathbf{V}_g = -\frac{v_g}{a}(\tan \phi + \cot \phi) \tag{7.31}$$

7.21 For the co-ordinate system of fig. 7.2 show that

$$\text{div } \mathbf{V} = \frac{\partial u}{\partial x} + \frac{\partial v}{\partial y} - \frac{v \tan \phi}{a} + \frac{\partial w}{\partial z} + \frac{2w}{a} \tag{7.32}$$

Show from typical values that the last term $2w/a$ may be neglected in comparison with $\partial w/\partial z$.

7.22 From (7.10) and (7.30) calculate an expression for the divergence of the geostrophic wind. Show by putting in typical values that the terms including $(\partial/\partial x)(1/\rho)$, $(\partial/\partial y)(1/\rho)$ are small compared with the others. Also show, again by inserting typical values, that div $\mathbf{V}_g$ is an order of magnitude smaller than either of the terms $\partial u/\partial x$, $\partial v/\partial y$.

7.23 From (7.1) and (7.18) show that an alternate form of the continuity equation is

$$\frac{d\rho}{dt} + \rho \text{ div } \mathbf{V} = 0 \tag{7.33}$$

7.24 Show that when the pressure p is used as the vertical co-ordinate, for a scalar ψ

$$\frac{d\psi}{dt} = \frac{\partial \psi}{\partial t} + u \frac{\partial \psi}{\partial x} + v \frac{\partial \psi}{\partial y} + \frac{dp}{dt} \cdot \frac{\partial \psi}{\partial p} \tag{7.34}$$

7.25 Starting with (7.18) and using the hydrostatic equation derive the continuity equation in isobaric co-ordinates (7.20) by algebraic manipulation.

7.26 Omitting the frictional force $\mathbf{F}$, multiply (7.4) by ρ and take its curl to obtain the *vorticity equation*

$$\mathbf{\nabla} \wedge \left(\rho \frac{d\mathbf{V}}{dt} \right) = \mathbf{\nabla} \wedge (2\rho \mathbf{V} \wedge \mathbf{\Omega}) - \mathbf{g} \wedge \mathbf{\nabla}\rho \tag{7.35}$$

Hence show that a state of hydrostatic equilibrium (i.e. $\mathbf{V} = 0$ everywhere) is impossible unless there are no variations of density on surfaces of constant geopotential. This result is known as the *Bjerknes–Jeffreys theorem*.

8
Atmospheric waves

8.1 Introduction

In chapter 7 we dealt with various approximations to the equations of motion particularly those applicable to the fairly large scale. No attempt was made to consider how different forms of motion might arise. Before approaching a discussion of why the circulation of the atmosphere is as we find it, it is instructive to look at simple but important types of wave motion which are present in the atmosphere. To isolate simple wave forms we shall again make severe approximations.

For each case the appropriate equations are (1) the momentum equations (§7.2), (2) the continuity equation (§7.7) and (3) the first law of thermodynamics. Our method of solution will be the perturbation method.

8.2 Sound waves

To illustrate the method we first consider sound waves. For a simple treatment of these it will suffice to assume motion along the x direction only and that the y and z components and gradients in these directions are zero. The Coriolis term and friction are also omitted and the motion is assumed adiabatic. The momentum equation (7.8) becomes

$$\frac{du}{dt} + \frac{1}{\rho}\frac{\partial p}{\partial x} = 0 \tag{8.1}$$

the continuity equation (7.18) becomes

$$\frac{d\rho}{dt} + \rho\frac{\partial u}{\partial x} = 0 \tag{8.2}$$

and the first law of thermodynamics for adiabatic motion is

$$p\rho^{-\gamma} = \text{constant} \tag{8.3}$$

where γ is the ratio of specific heats of dry air. Combining (8.2) and (8.3) we obtain

$$\frac{1}{\gamma}\frac{d\ln p}{dt} + \frac{\partial u}{\partial x} = 0 \tag{8.4}$$

87

The perturbation method consists of allowing the variables to be written as the sum of a mean component (represented by a bar) and a variable component (represented by a prime)

i.e

$$u = \bar{u} + u'$$

$$p = \bar{p} + p' \qquad (8.5)$$

$$\rho = \bar{\rho} + \rho'$$

Here $p' \ll \bar{p}, \rho' \ll \bar{\rho}$ but u' is not necessarily $< \bar{u}$ as the solutions are still valid if $\bar{u} = 0$. Substituting in (8.1) and (8.4), neglecting products of primed quantities, assuming that differentials of mean quantities are zero, and also using (7.1) we find

$$\left(\frac{\partial}{\partial t} + \bar{u} \frac{\partial}{\partial x} \right) u' + \frac{1}{\bar{\rho}} \frac{\partial p'}{\partial x} = 0 \qquad (8.6)$$

$$\left(\frac{\partial}{\partial t} + \bar{u} \frac{\partial}{\partial x} \right) p' + \gamma \bar{p} \frac{\partial u'}{\partial x} = 0 \qquad (8.7)$$

Eliminating u' between these equations

$$\left(\frac{\partial}{\partial t} + \bar{u} \frac{\partial}{\partial x} \right)^2 p' - \frac{\gamma \bar{p}}{\bar{\rho}} \frac{\partial^2 p'}{\partial x^2} = 0 \qquad (8.8)$$

which is a wave equation having solutions

$$p' = \text{Re} \{A \exp ik(x - ct)\} \qquad (8.9)$$

with a constant A and

$$c = \bar{u} \pm (\gamma \bar{p}/\bar{\rho})^{1/2} \qquad (8.10)$$

The speed of sound waves relative to the flow $\bar{u}$ is therefore $(\gamma \bar{p}/\bar{\rho})^{1/2}$.

8.3 Gravity waves

Since sound waves are longitudinal, only one dimension need be considered to obtain the sound wave solution. Other waves exist where the oscillation is transverse to the direction of propagation; some of these waves are known as *gravity waves*. Their existence is illustrated by the discussion of §1.4 where we considered the stability of a parcel displaced vertically from its equilibrium level. In problem 1.10 for a stable stratification the frequency of oscillation (the Brunt–Vaisala frequency) of such a parcel was found on the assumption that the environment is unaffected by the motion – a simple calculation which applies in the limiting case of large vertical scale and small horizontal scale.

To obtain a reasonably simple solution to the basic equations for gravity waves we shall (1) work in two dimensions and ignore motion or gradients along the y direction, (2) assume that the horizontal scale of the wave is sufficiently small that the Coriolis term may be neglected by comparison with the other terms. (3) ignore the friction term, (4) assume that the unperturbed atmosphere is at rest. The components of the momentum equation (7.4) are

$$\frac{du}{dt} + \frac{1}{\rho}\frac{\partial p}{\partial x} = 0 \tag{8.11}$$

$$\frac{dw}{dt} + \frac{1}{\rho}\frac{\partial p}{\partial z} + g = 0 \tag{8.12}$$

The continuity equation (7.33) becomes

$$\frac{1}{\rho}\frac{d\rho}{dt} + \frac{\partial u}{\partial x} + \frac{\partial w}{\partial z} = 0 \tag{8.13}$$

and the first law of thermodynamics for adiabatic motion means that the potential temperature θ remains constant, i.e.

$$\frac{d \ln \theta}{dt} = 0 \tag{8.14}$$

From an equation similar to (3.3), θ can be expressed in terms of ρ and p such that

$$\frac{d \ln \theta}{dt} = \frac{1}{\gamma}\frac{d \ln p}{dt} - \frac{d \ln \rho}{dt} \tag{8.15}$$

For the unperturbed atmosphere neither θ, p nor ρ vary in the horizontal.

Let perturbations be introduced in (8.11), (8.12), (8.13) and (8.15), such that $u = u'$, $w = w'$, $\rho = \bar{\rho} + \rho'$, and $p = \bar{p} + p'$. Then using (7.1), the hydrostatic equation $\bar{p}^{-1}(\partial \bar{p}/\partial z) = -H^{-1}$, the equation of state $\bar{p}/\bar{\rho} = gH$, where H is the scale height, also neglecting products of primed quantities, the four equations (8.11), (8.12), (8.13) and (8.15) become for the unknowns u', w', $\rho'/\bar{\rho}$ and $p'/\bar{p}$ (problem 8.2)

$$\frac{\partial u'}{\partial t} + gH\frac{\partial}{\partial x}\left(\frac{p'}{\bar{p}}\right) = 0 \tag{8.16}$$

$$\frac{\partial w'}{\partial t} + g\left(\frac{\rho'}{\bar{\rho}}\right) + gH\frac{\partial}{\partial z}\left(\frac{p'}{\bar{p}}\right) - g\left(\frac{p'}{\bar{p}}\right) = 0 \tag{8.17}$$

$$\frac{\partial u'}{\partial x} + \frac{\partial w'}{\partial z} - \frac{w'}{H} + \frac{\partial}{\partial t}\left(\frac{\rho'}{\bar{\rho}}\right) = 0 \tag{8.18}$$

$$-Bw' + \frac{\partial}{\partial t}\left(\frac{\rho'}{\bar{\rho}}\right) - \frac{1}{\gamma}\frac{\partial}{\partial t}\left(\frac{p'}{\bar{p}}\right) = 0 \tag{8.19}$$

where, according to the nomenclature of problem 1.10, the vertical stability parameter B has been written for $\partial \ln \theta / \partial z$.

89

In solving these equations we shall assume first an isothermal atmosphere in which H is a constant, in which case $B = (\gamma - 1)/\gamma H$ and is, of course, also constant. We look for wave solutions such that each of the four unknowns vary as

$$\exp(\alpha z) \exp i(\omega t + kx + mz)$$

in which the first exponential has been introduced to allow for variation of amplitude with altitude. Solutions of this kind will exist if after substituting them into (8.16) to (8.19) the determinant of the coefficients is equal to zero. On equating imaginary parts we find that unless $m = 0$, $\alpha = 1/2H$. Solutions with the first condition have no phase variation in the vertical and are known as *external waves*; the most important of these are *surface waves* whose energy is concentrated at a boundary or discontinuity in the atmosphere similar to surface waves on the ocean. When $m \neq 0$, we have *internal waves*. Substituting $\alpha = 1/2H$ into the equation and equating the real part of the determinant of the coefficients to zero the following dispersion relation results (problem 8.3):

$$m^2 = k^2 \left(\frac{\omega_B^2}{\omega^2} - 1 \right) + \frac{(\omega^2 - \omega_a^2)}{c^2} \tag{8.20}$$

where c is the velocity of sound,

$$\omega_B^2 = \frac{(\gamma - 1)g}{\gamma H} \tag{8.21}$$

is the square of the Brunt–Vaisala frequency for the isothermal atmosphere (problem 8.4) and

$$\omega_a = \frac{1}{2} \left(\frac{\gamma g}{H} \right)^{1/2} = \frac{c}{2H} \tag{8.22}$$

is known as the *acoustic cut-off frequency*.

The dispersion relation (8.20) is illustrated in fig. 8.1 where curves of $m = 0$ are plotted on a ω–k diagram. Two regions where m^2 is positive are apparent, the higher frequency region ($\omega > \omega_a$ when $k = 0$) describes acoustic waves and the lower frequency region, gravity waves. Since $\omega_B < \omega_a$ (problem 8.6) these two regions are well separated. Note also from (8.20) that if $m \ll k$, i.e. for deep waves of short horizontal wavelength compared with vertical wavelength, $\omega \simeq \omega_B$ as would be expected from our simple derivation of ω_B in problem 1.10.

If the assumption that the atmosphere is isothermal is removed, (8.20) may still be employed as a reasonable approximation with ω_B as the Brunt–Vaisala frequency for the atmosphere in question, provided $\omega_B < \omega_a$ as is normally the case throughout the atmosphere (problem 8.6).

If the atmosphere is in a state of uniform zonal flow $\bar{u}$, (8.20) still holds provided ω is replaced by $\omega + \bar{u}k$, the frequency which would be seen by an

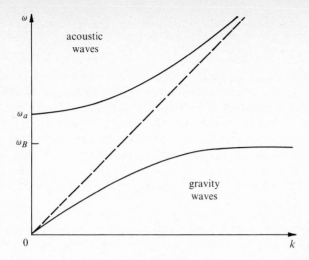

Fig. 8.1. Dispersion curves for gravity waves. The full lines corre-
spond to $m = 0$; the dashed line is for a wave having the velocity of
sound c.

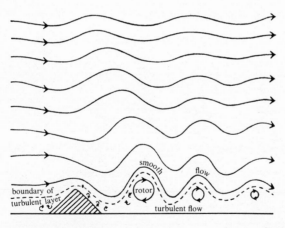

Fig. 8.2. The structure of lee waves according to Gerbier & Berenger
(1961).

observer moving with the basic flow and known as the *intrinsic frequency*.

Equation (8.20) applies to waves having a wide range of wavelength, frequency and velocity. For gravity waves of horizontal wavelength of the order
of a few kilometres, the first term is much larger than the second. In this case,
allowing also for the presence of a uniform zonal wind $\bar{u}$ the dispersion relation (8.20) may be written

$$\frac{(\omega - \bar{u}k)^2}{\omega_B^2} = \frac{k^2}{m^2 + k^2} \tag{8.23}$$

The most easily observed example of such gravity waves is that of waves in the lee of mountains. Air which is forced to flow over a mountain under stable conditions is set into a gravity-wave oscillation as it moves downstream from the mountain. If the amplitude is large enough and the conditions of temperature and humidity suitable, clouds will form in the regions where the air has been lifted. Such waves will, of course, be stationary with respect to the mountain and the horizontal component of phase velocity relative to the surface will be zero, in which case for $k \ll m$ from (8.23) we have

$$m = \omega_B/\bar{u} \tag{8.24}$$

Since the energy source is near the surface, energy is propagated upwards. Their group velocity is, therefore, upwards and their phase velocity downwards (problem 8.10). Consequently the lines of constant phase in the stationary waves tilt with height backwards relative to the mean flow (fig. 8.2), although in practice, because the mean wind varies with height, the situation is rarely so simple as this.

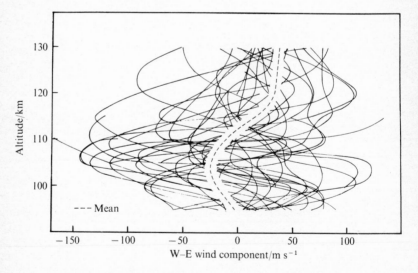

Fig. 8.3. Profiles of the W–E component of the wind as measured from 36 rockets which carried chemical trail experiments all near 30°N latitude between 15 November and 14 December or near 30°S latitude between 15 May and 14 June (from COSPAR 1972).

Gravity waves represent only a minor component of the motion in the lower atmosphere. Above 75 km, however, atmospheric motion is probably dominated by them (fig. 8.3). Gravity waves include tidal motions, tides being a special case of gravity waves having a particular horizontal scale and a particular period. These motions show up for instance in the variations of ionized layers as well as in winds determined from the movement of trails left by rockets or by meteors. Although mainly generated in the lower atmosphere gravity waves are propagated upwards, where as the density decreases their amplitude increases (problem 8.8). At levels above 75 km or so they are dissipated by viscous damping. Because of this they contribute significantly to the energy budget of the upper mesosphere and lower thermosphere.

8.4 Rossby waves

The very large scale wave features which are observed in the flow on the planetary scale are known as *planetary waves* (fig. 8.4). In their simplest form they occur because of the variation of the Coriolis parameter with latitude and are known as *Rossby waves*.

The simplest Rossby wave solution is obtained for an atmosphere of constant density under the assumption of uniform zonal flow $\bar{u}$ and no vertical motion. The appropriate momentum equations (7.8) and the continuity equation (7.19) are

$$\frac{du}{dt} + \frac{1}{\rho}\frac{\partial p}{\partial x} - fv = 0 \tag{8.25}$$

$$\frac{dv}{dt} + \frac{1}{\rho}\frac{\partial p}{\partial y} + fu = 0 \tag{8.26}$$

$$\frac{\partial u}{\partial x} + \frac{\partial v}{\partial y} = 0 \tag{8.27}$$

Operating on (8.25) with $\partial/\partial y$ and (8.26) with $\partial/\partial x$ and subtracting

$$\frac{d}{dt}\left(\frac{\partial v}{\partial x} - \frac{\partial u}{\partial y}\right) + f\left(\frac{\partial v}{\partial y} + \frac{\partial u}{\partial x}\right) + v\frac{\partial f}{\partial y} = 0 \tag{8.28}$$

The quantity $(\partial v/\partial x) - (\partial u/\partial y)$ is one component of curl $\mathbf{V}$ which is known as the *vorticity* ζ; it may be considered as a vector equal to twice the local angular velocity of fluid elements. The second term in (8.28) is zero by (8.27). Because the Coriolis parameter f only varies with latitude it is possible to write (8.28) in the form

$$\frac{d(\zeta + f)}{dt} = 0 \tag{8.29}$$

The quantity $\zeta + f$ is known as the *absolute vorticity*; it is the vorticity due to the rotation of the fluid itself combined with that due to the earth's rotation. Equation (8.29) shows that under the conditions we have imposed of

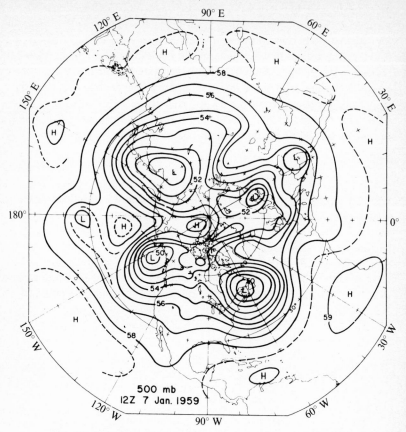

Fig. 8.4. Geopotential height of the 500 mb pressure surface in hundreds of metres for the northern hemisphere for 7 January 1959, 1200 Z (from Palmén & Newton, 1969). The large scale waves are known as *planetary waves*.

non-divergent, frictionless flow, absolute vorticity is conserved.

To solve (8.29) we assume a linear relation between f and y, i.e. we let $f = f_0 + \beta y$ where β is a constant (this is known as the *β-plane approximation*). We assume a uniform zonal flow $\bar{u}$ in the unperturbed situation and introduce perturbations, i.e. $u = \bar{u} + u'$, $v = v'$ so that (8.29) becomes

$$\left(\frac{\partial}{\partial t} + \bar{u}\frac{\partial}{\partial x}\right)\left(\frac{\partial v'}{\partial x} - \frac{\partial u'}{\partial y}\right) + \beta v' = 0 \tag{8.30}$$

Because the flow is non-divergent in the horizontal a *stream function* ψ can be introduced such that by using it (8.27) is automatically satisfied,

i.e. $$u' = -\frac{\partial \psi}{\partial y}, \qquad v' = \frac{\partial \psi}{\partial x} \tag{8.31}$$

Atmospheric waves

Substituting in (8.30)

$$\left(\frac{\partial}{\partial t} + \bar{u}\frac{\partial}{\partial x}\right)\nabla^2\psi + \beta\frac{\partial\psi}{\partial x} = 0 \tag{8.32}$$

For wave solutions $\psi = \mathrm{Re}\{\psi_0 \exp i(\omega t + kx + ly)\}$ to be possible the dispersion relation

$$c = -\frac{\omega}{k} = \bar{u} - \frac{\beta}{k^2 + l^2} \tag{8.33}$$

must be satisfied. The velocity relative to the zonal flow is $c - \bar{u}$ where c is the phase velocity in the x direction. Rossby waves, therefore, drift to the west relative to the basic flow, at typical speeds of a few metres per second (problem 8.14). Note that the phase speed of the waves increases with wavelength.

8.5 The vorticity equation

Under the assumptions of two dimensional flow in an atmosphere of uniform density (8.29) is a statement of the conservation of absolute vorticity. Before dealing with the propagation of Rossby-type waves in three dimensions we need to find an equation to describe vorticity changes under less stringent assumptions. For motions of synoptic scale this can be done by starting with the approximate horizontal momentum equations (8.25) and (8.26). Operating on (8.25) with $\partial/\partial y$ and (8.26) with $\partial/\partial x$ and subtracting, also noting that $df/dt = v\,\partial f/\partial y$ we have for the rate of change of absolute vorticity

$$\frac{d}{dt}(\zeta + f) = -(\zeta + f)\left(\frac{\partial u}{\partial x} + \frac{\partial v}{\partial y}\right)$$
$$-\left(\frac{\partial w}{\partial x}\frac{\partial v}{\partial z} - \frac{\partial w}{\partial y}\frac{\partial u}{\partial z}\right) + \frac{1}{\rho^2}\left(\frac{\partial\rho}{\partial x}\frac{\partial p}{\partial y} - \frac{\partial\rho}{\partial y}\frac{\partial p}{\partial x}\right) \tag{8.34}$$

Equation (8.34) is known as the *vorticity equation* (cf (7.35)). The first term on the r.h.s. is the most important; it arises because of the horizontal divergence. If there is positive horizontal divergence, air is flowing out of the region in question, and the vorticity decreases. This is the same effect as occurs with a rotating body whose angular velocity decreases because of angular momentum conservation if its moment of inertia increases.

Scale analysis of (8.34) shows that for synoptic scale motions the last two terms are considerably smaller than the others (problem 8.15) and that to a first approximation

$$\frac{d_h}{dt}(\zeta + f) = -(\zeta + f)\left(\frac{\partial u}{\partial x} + \frac{\partial v}{\partial y}\right) \tag{8.35}$$

where d_h/dt denotes $\partial/\partial x + \partial/\partial y$.

95

It is instructive to apply (8.35) to an atmosphere of constant density and temperature for which the continuity equation is (7.19) so that (8.35) becomes

$$\frac{d_h}{dt}(\zeta + f) = (\zeta + f)\frac{\partial w}{\partial z} \tag{8.36}$$

Because of the constant temperature the geostrophic wind is independent of the height z. Further, because to a first approximation the vorticity is equal to the vorticity of the geostrophic wind the vorticity will not vary with height (problem 8.16). Therefore, integrating (8.36) between levels z_1 and z_2 where $z_2 - z_1 = h$ we have

$$\frac{1}{(\zeta + f)}\frac{d_h}{dt}(\zeta + f) = \frac{w(z_2) - w(z_1)}{h} \tag{8.37}$$

Now considering the fluid which at one time is confined between the levels distance h apart, we have

$$\frac{dh}{dt} = w(z_2) - w(z_1) \tag{8.38}$$

so that (8.37) may now be written

$$\frac{d_h}{dt}\left(\frac{\zeta + f}{h}\right) = 0 \tag{8.39}$$

Equation (8.39) is a simplified statement of the conservation of *potential vorticity*. It has important consequences for atmospheric flow. Consider, for instance, adiabatic flow over a mountain barrier. As a column of air flows over the barrier its vertical extent decreases so that ζ must also decrease. A westward moving airstream will, therefore, move equatorwards as it passes over the barrier (problem 8.17).

8.6 Three dimensional Rossby-type waves

In §8.4 we assumed two dimensional structure only for the Rossby wave solutions. To find how such waves propagate vertically we need to look for the constraints imposed on three dimensional solutions. To simplify the treatment we still wish to work with an atmosphere at nearly constant density and so we employ the *Boussinesq approximation* which allows us to neglect changes of density except where they are coupled with gravity to produce buoyancy forces. With this approximation the equation of continuity is as for an incompressible fluid, i.e. (7.19) applies, so that the appropriate vorticity equation is (8.36). With a uniform unperturbed zonal flow $\bar{u}$, the perturbation form of (8.36) is

$$\left(\frac{\partial}{\partial t} + \bar{u}\frac{\partial}{\partial x}\right)\zeta' + v'\frac{\partial f}{\partial y} - f\frac{\partial w'}{\partial z} = 0 \tag{8.40}$$

where ζ has been neglected compared with f in the r.h.s. of (8.36) (problem 8.16).

We also require the perturbation form of the hydrostatic equation

$$\frac{1}{\bar{\rho}}\frac{\partial p'}{\partial z} + \frac{\rho'}{\bar{\rho}}g = 0 \tag{8.41}$$

and the thermodynamic equation which under the Boussinesq approximation becomes

$$\left(\frac{\partial}{\partial t} + \bar{u}\frac{\partial}{\partial x}\right)\frac{\rho'}{\bar{\rho}} - Bw' = 0 \tag{8.42}$$

Comparing (8.42) with (8.19), note that since under the approximation of a nearly incompressible fluid, $\gamma \to \infty$, the last term in (8.19) may be neglected.

Eliminating ρ' and w' from (8.40), (8.41) and (8.42) and putting $\beta = \partial f/\partial y$ (the β-plane approximation) we have

$$\left(\frac{\partial}{\partial t} + \bar{u}\frac{\partial}{\partial x}\right)\left(\zeta' + \frac{f_0}{gB\bar{\rho}}\frac{\partial^2 p'}{\partial z^2}\right) + \beta v' = 0 \tag{8.43}$$

Now most of the vorticity perturbation arises from the vorticity perturbation of the geostrophic wind (problem 8.16) whose components u_g' and v_g' are

$$u_g' = -\frac{1}{f_0\bar{\rho}}\frac{\partial p'}{\partial y}, \qquad v_g' = \frac{1}{f_0\bar{\rho}}\frac{\partial p'}{\partial x} \tag{8.44}$$

For the geostrophic wind, as in (8.31), we can define a stream function ψ which comparing with (8.44) will be equal to $p'/f_0\bar{\rho}$. Introducing this into (8.43) and also approximating v' by its geostrophic value in the last term, (8.43) becomes

$$\left(\frac{\partial}{\partial t} + \bar{u}\frac{\partial}{\partial x}\right)\left(\nabla^2\psi + \frac{f_0^2}{gB}\frac{\partial^2\psi}{\partial z^2}\right) + \beta\frac{\partial\psi}{\partial x} = 0 \tag{8.45}$$

These last substitutions involve the *quasi-geostrophic approximation* in which the wind is replaced by its geostrophic value except in the divergence term. For wave solutions

$$\psi = \text{Re}\{\psi_0 \exp i(\omega t + kx + ly + mz)\}$$

to be possible for (8.45) the dispersion relation

$$c = -\frac{\omega}{k} = \bar{u} - \frac{\beta}{(k^2 + l^2 + m^2 f_0^2/gB)} \tag{8.46}$$

must be satisfied. Since $f_0^2/gB \ll 1$ (problem 8.18), vertical wavelengths are of the order of 1% of horizontal wavelengths.

We might expect some of these large scale waves to be forced by features at the surface, by mountains or large land masses, in which case they would be

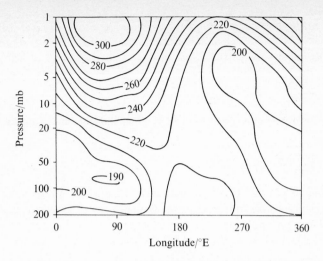

Fig. 8.5. Temperature cross-section (in K) of the stratosphere around the latitude circle 65° N as derived from the Selective Chopper Radiometer on Nimbus 4 for 4 January 1971 at the peak of a *stratospheric warming*. Note the large amplitude wavenumber one (the wavenumber refers to the number of complete cycles of a variation which occur around a latitude circle) and the westward tilt of the wave with height. (After Houghton, 1972)

stationary with respect to the surface. For c to be zero from (8.46)

$$m^2 = \frac{gB}{f_0^2}\left\{\frac{\beta}{\bar{u}} - (k^2 + l^2)\right\}\tag{8.47}$$

For vertical propagation to occur m^2 must be positive, which sets the condition

$$0 < \bar{u} < \beta/(k^2 + l^2)\tag{8.48}$$

Under conditions of easterly flow, therefore, or very large westerly flow no vertical propagation will occur. This result, first obtained by Charney & Drazin (1961), is confirmed by the lack of planetary wave activity in the stratosphere during the summer months when the mean stratospheric flow is easterly. Further the waves of largest horizontal wavelength propagate most readily. Fig. 8.5 and 8.6 illustrate some such waves observed in the stratosphere; where their amplitude is sufficiently large they are known as *stratospheric warmings*.

Problems

8.1 Write down the horizontal equations of motion for an atmosphere in which pressure oscillations are negligible and in which there is no basic flow, i.e.

$$\left.\begin{array}{c}\dfrac{\partial u}{\partial t} - fv = 0 \\[2ex] \dfrac{\partial v}{\partial t} + fu = 0 \end{array}\right\} \qquad (8.49)$$

Substituting $V = u + iv$ ($i = \sqrt{-1}$) solve the resulting equation on the assumption that f is constant and show that *inertial oscillations* occur with a period $2\pi/f$ (cf. also problem 7.6).

8.2   Derive (8.16) to (8.19) from (8.11) to (8.15).

8.3   Derive (8.20)

8.4   Show that for an isothermal atmosphere

$$B = (\gamma - 1)/\gamma H \qquad (8.50)$$

From the result of problem 1.10 find ω_B for an isothermal atmosphere.

8.5   With $\bar{u} = 0$, from (8.23), find the horizontal phase velocity of gravity waves for various values of horizontal and vertical wavelength.

8.6   Show that ω_a defined by (8.22) is always greater than ω_B for an isothermal atmosphere defined by (8.21). Calculate the value of the lapse rate of temperature with altitude for which the Brunt–Vaisala frequency is equal to ω_a.

8.7   Consider a pattern of motion in which fluid elements move at an angle θ to the vertical. Show that for an atmosphere where the Brunt–Vaisala frequency is ω_B the elements will oscillate with a frequency $\omega = \omega_B \cos \theta$.

Now (following Lindzen, 1971) imagine waves propagating upwards at angle θ to the vertical excited by a surface with uniform corrugations distant $2\pi/k$ apart moving horizontally at the base of the atmosphere. Show that the ratio of horizontal wavelength ($2\pi/k$) to vertical wavelength ($2\pi/m$) is $\tan \theta$. Hence deduce the relation (cf. (8.20))

$$m^2 = \frac{1 - (\omega/\omega_B)^2}{(\omega/\omega_B)^2} k^2$$

8.8   Show that in a wave which is propagating upwards with constant kinetic energy the velocity in the wave will vary with altitude approximately as $\exp(z/2H)$ where H is the scale height. If a gravity wave at 100 km has an amplitude of 100 m s^{-1}, what would be its amplitude at the surface where it originates?

8.9   In (8.24) with $\bar{u} = 10 \text{ m s}^{-1}$, calculate the vertical wavelength for (1) an isothermal atmosphere, (2) an atmosphere with a lapse rate of 5 K km^{-1}.

8.10   For gravity waves in an atmosphere with no mean flow ($\bar{u} = 0$) and having a dispersion relation given by (8.23) compute the horizontal and vertical components of the group velocity. Show that for $k \ll m$,

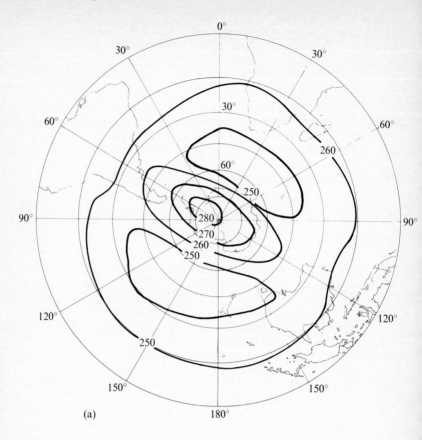

(a)

the vertical component of group velocity is equal and opposite to the vertical component of phase velocity.

8.11   Write down the horizontal momentum equations for perturbations in an atmosphere with no mean flow ($\bar{u} = 0$), i.e.

$$\frac{\partial u'}{\partial t} + \frac{1}{\bar{\rho}} \frac{\partial p'}{\partial x} - fv' = 0$$

$$\frac{\partial v'}{\partial t} + \frac{1}{\bar{\rho}} \frac{\partial p'}{\partial y} + fu' = 0 \qquad (8.51)$$

Add (8.17) to (8.19) in their simplified form, assume no variation

100

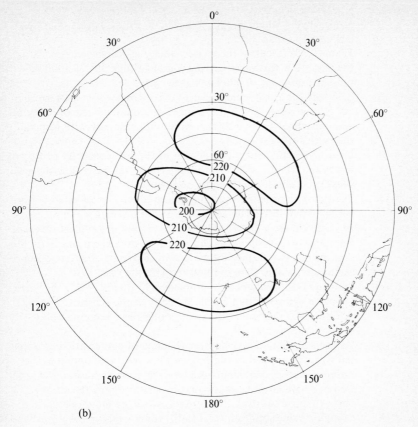

(b)

Fig. 8.6. Illustrating a quasi-stationary wave in southern hemisphere stratosphere of wavenumber 2 as observed by Selective Chopper Radiometer on Nimbus 4 satellite on 25 September 1971 (after Harwood, 1975). The quantity plotted is temperature in K of a layer ~ 10 km thick centred at (a) ~ 45 km altitude and (b) 15 km altitude. Note that there is about half a wavelength of the wave in the vertical between 45 and 15 km.

with y and plane wave solutions as in §8.3. Show that the equations are satisfied and that the appropriate dispersion relation is now

$$\omega^2 - f^2 = \frac{k^2}{m^2} (\omega_B^2 - \omega^2) \tag{8.52}$$

showing that ω is always $> f$.

This result shows the constraint on ω for gravity waves which are of sufficiently large horizontal extent for the Coriolis term to be important. It is particularly important for consideration of tides. Because f varies from zero at the equator to 2π (12 hours)$^{-1}$ at the

101

poles, internal gravity waves with periods shorter than 12 hours can propagate anywhere but the diurnal tide will be largely confined to latitudes less than 30°.

8.12 From (8.33) calculate the phase velocity of Rossby waves relative to the basic flow for waves moving round latitude 30° of 3000 km latitudinal width and 10 000 km wavelength.

8.13 Calculate the zonal wind under which a Rossby wave pattern at 60° N of wavenumber 3 (i.e. 3 maxima around a circle of longitude) and latitudinal width 3000 km would be stationary with respect to the earth's surface.

8.14 From (8.33) find the x component of the group velocity of Rossby waves and when $k > l$ show that relative to the basic flow the group velocity is opposite to the phase velocity.

8.15 For synoptic scale motions in mid latitudes a typical value of horizontal dimension is 1000 km, of horizontal velocity 10 m s^{-1}, of vertical velocity 1 cm s^{-1}. What are corresponding typical values of fractional density variations $\delta\rho/\rho$ and pressure variations $\delta p/p$. Show that the last two terms on the r.h.s. of (8.34) are about an order of magnitude smaller than the first term on the r.h.s.

8.16 Show by scale analysis as in problem 8.15 that in (8.35) the vorticity ζ may be approximated by the vorticity of the geostrophic wind.

Show also that under the same conditions ζ is considerably smaller than f.

8.17 Consider the implications of the theorem of conservation of potential vorticity applied to the easterly flow over a mountain barrier. A difference arises compared with westerly flow because of the variation of f with latitude. Show that for easterly flow the air must begin to move southward before it reaches the mountain barrier.

8.18 Compute values of f^2/gB (cf. (8.46)) for several typical atmospheric stabilities (e.g. for an isothermal atmosphere and for an atmosphere having a lapse rate of 5 K km^{-1}).

8.19 Compute the vertical phase velocity and group velocity for waves described by (8.46). Show that for waves in which energy is propagated upwards the phase velocity is downwards and hence that such waves slope towards the west with increasing altitude.

8.20 Write down the horizontal momentum equations (8.51) for perturbations u', v', p' in a form applicable to an equatorial β-plane, i.e. with $f = \beta y$. Assume a solution with $v' = 0$, and

$$u' = \text{Re}\{u_0' \exp i\,(\omega t - kx)\}$$

Show that in this case the amplitude u_0' varies with latitude as

$$\exp\left(\frac{-\beta y^2 k}{2\omega}\right)$$

Note that for a satisfactory solution k must be positive, i.e. the wave must be *eastward moving*. Plot the pressure field associated with the variation in zonal velocity u'. These are *Kelvin waves*; they have been observed in the equatorial stratosphere.

102

9
Turbulence

9.1 The Reynolds Number

In writing down the equations of motion (§7.2) a term F was included to allow for friction. How to deal with this term is the subject of this chapter.

To begin with, consider fluid flow along a pipe. When all elements of fluid are moving along the direction of the fluid's mean motion the flow is *laminar* and internal friction in the fluid will be due only to molecular viscosity. When, however, individual elements are moving irregularly compared with the mean motion the flow is *turbulent*.

Reynolds' classical experiments on the flow of fluid of uniform density through pipes showed that a non-dimensional quantity could be found whose approximate value determined whether the sheared flow was laminar or turbulent. This quantity is known as the Reynolds number Re; for a fluid of kinematic viscosity* ν moving with velocity V it is given by

$$Re = LV/\nu$$

where L is a typical length scale of the motion (e.g. the diameter of the pipe). For large values of Re greater than about 6000, turbulent flow occurs, for smaller values the flow is laminar. For air at STP $\nu \simeq 1.5 \times 10^{-5}\,\mathrm{m^2\,s^{-1}}$ so that if $L = 1$ m, provided $V > \sim 0.1\,\mathrm{m\,s^{-1}}$ the condition that $Re > 6000$ is satisfied. With typical atmospheric scales and velocities, therefore, we expect the flow to be turbulent.

It is useful in any consideration of a particular type of atmospheric motion to restrict the consideration to a particular scale; in the last chapter this was done in order to isolate particular types of wave motion. The fact of turbulence means that this isolation can never in fact be complete. Motion on all smaller scales than the one being considered will exist; interactions and exchange of energy will take place between motion on these different scales. One of the main objects, therefore, of the theory of turbulence is to provide means whereby, for any particular problem, the effect of turbulent motions on smaller scales than the one being considered can be expressed in terms of the parameters of the mean flow.

In this chapter we shall first describe the effect of turbulence in the

---

* The kinematic viscosity is the viscosity divided by the density — its units are $\mathrm{m^2\,s^{-1}}$.

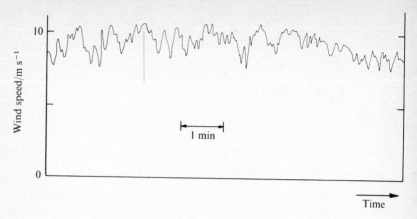

Fig. 9.1. Typical trace of wind speed taken at a height of 10 m over Lake Ontario.

boundary layer close to the earth's surface and then discuss in rather general terms the interactions between motion on various scales in the atmosphere at large.

9.2 Reynolds stresses

At any given point and time the components of velocity u, v and w may be expressed in terms of mean components (denoted by bars) and fluctuating components (denoted by primes), e.g. $u = \bar{u} + u'$. For this to make sense, a suitable averaging period has to be found such that the mean quantity is substantially independent of the precise averaging period. This period will depend on the particular problem under consideration; for motion in the boundary layer, 5 or 10 min is a suitable value (figs. 9.1 and 9.4).

The horizontal momentum equations (7.8) neglecting friction when due only to viscosity are

$$\frac{\partial u}{\partial t} + u \frac{\partial u}{\partial x} + v \frac{\partial u}{\partial y} + w \frac{\partial u}{\partial z} - fv + \frac{1}{\rho} \frac{\partial p}{\partial x} = 0 \tag{9.1}$$

$$\frac{\partial v}{\partial t} + u \frac{\partial v}{\partial x} + v \frac{\partial v}{\partial y} + w \frac{\partial v}{\partial z} + fu + \frac{1}{\rho} \frac{\partial p}{\partial y} = 0 \tag{9.2}$$

and the continuity equation (7.18) is

$$\frac{\partial}{\partial x} (\rho u) + \frac{\partial}{\partial y} (\rho v) + \frac{\partial}{\partial z} (\rho w) + \frac{\partial \rho}{\partial t} = 0 \tag{9.3}$$

Multiplying (9.1) by ρ and (9.3) by u and adding we have

104

$$\frac{\partial}{\partial t}(\rho u) + \frac{\partial}{\partial x}(\rho u^2) + \frac{\partial}{\partial y}(\rho uv) + \frac{\partial}{\partial z}(\rho uw) - f\rho v + \frac{\partial p}{\partial x} = 0 \qquad (9.4)$$

A similar equation results from combining (9.2) and (9.3) in the same way.

We now substitute $u = \bar{u} + u'$, $v = \bar{v} + v'$ and $w = \bar{w} + w'$ into (9.4), remembering that, since $\overline{u'}, \overline{v'}, \overline{w'} = 0$ by definition, two terms only result from averaging the product of varying quantities, e.g. $\overline{uw} = \overline{(\bar{u} + u')(\bar{w} + w')} = \bar{u}\bar{w} + \overline{u'w'}$. Since relative variations in density are very much less than those in velocity, fluctuations in the density will be neglected. The resulting equation

$$\frac{\partial \bar{u}}{\partial t} + \bar{u}\frac{\partial \bar{u}}{\partial x} + \bar{v}\frac{\partial \bar{u}}{\partial y} + \bar{w}\frac{\partial \bar{u}}{\partial z} - f\bar{v} + \frac{1}{\rho}\frac{\partial p}{\partial x}$$

$$+ \frac{1}{\rho}\left[\frac{\partial}{\partial x}\rho\overline{u'u'} + \frac{\partial}{\partial y}\rho\overline{u'v'} + \frac{\partial}{\partial z}\rho\overline{u'w'}\right] = 0 \qquad (9.5)$$

contains all the terms of (9.1) now expressed in terms of the mean flow together with terms which express the effect of the fluctuating components; these terms describe what are known as the *Reynolds stresses* or the *eddy stresses*; they represent the flux of momentum due to the turbulent or eddy motions (problem 9.1).

The formulation of (9.5) suggests ways in which the Reynolds stresses might be measured, but gives no indication of how to express them in terms of the mean quantities. The simplest approach is to draw an analogy with molecular viscosity and considering a plane boundary in the xy-plane write for the eddy stress in the x direction on a plane parallel to the boundary

$$-\rho\overline{u'w'} = \rho K\frac{\partial \bar{u}}{\partial z} \qquad (9.6)$$

where K is the *coefficient of eddy viscosity* (with the same dimensions as kinematic viscosity) and is effectively defined by (9.6). Typical atmospheric values of K lie in the range $1-100\,\mathrm{m^2\,s^{-1}}$. These are high values when compared with the molecular viscosity of ordinary fluids (typically $10^{-5}\,\mathrm{m^2\,s^{-1}}$ for gases at STP). They approximate to the viscosity of thick treacle and demonstrate the effectiveness of eddy motions compared with molecular motions in transferring momentum.

9.3 Eckman's solution

Turbulent transfer is the main mechanism by which heat, moisture and momentum are exchanged between the atmosphere and the earth's surface. The region where these transfers occur is the *boundary layer*. First consider momentum transfer in this layer and let us assume a situation in which there is a shear of the mean wind in the vertical direction only. Neglecting, therefore, in (9.5) the stress terms involving horizontal gradients, we express the vertical eddy stress term (i.e. the third one) as in (9.6) with K as a constant. For a situation where the wind above the boundary layer is geostrophic and where horizontal temperature gradients through the boundary layer are negligible (i.e. the geostrophic wind through the layer does not change with height), the following equations describe the motion in the boundary layer

$$\frac{1}{\rho}\frac{\partial p}{\partial x} - fv - K\frac{\partial^2 u}{\partial z^2} = 0 \tag{9.7}$$

$$\frac{1}{\rho}\frac{\partial p}{\partial y} + fu - K\frac{\partial^2 v}{\partial z^2} = 0 \tag{9.8}$$

The components of the geostrophic wind (7.10) are

$$u_g = -\frac{1}{\rho f}\frac{\partial p}{\partial y}, \qquad v_g = \frac{1}{\rho f}\frac{\partial p}{\partial x} \tag{9.9}$$

Substituting these in (9.7) and (9.8), multiplying (9.8) by $i = \sqrt{-1}$, and adding:

$$K\frac{\partial^2 (u + iv)}{\partial z^2} - if(u + iv) + if(u_g + iv_g) = 0 \tag{9.10}$$

Suitable boundary conditions are zero wind components at the surface, i.e. $u = v = 0$ at $z = 0$ and geostrophic wind at high levels. For simplicity, assume that this geostrophic wind is zonal, i.e. $v_g = 0$, so that $u \to u_g$, $v \to 0$ as $z \to \infty$ is the other boundary condition.

The solution of (9.10) is

$$u + iv = u_g\left\{1 - \exp\left[-\left(\frac{f}{2K}\right)^{1/2}(1 + i)z\right]\right\} \tag{9.11}$$

or in components

$$u = u_g[1 - \exp(-\gamma z)\cos \gamma z] \tag{9.12}$$

$$v = u_g \exp(-\gamma z)\sin \gamma z \tag{9.13}$$

where $\gamma = \left(\frac{f}{2K}\right)^{1/2}$

Equations (9.12) and (9.13) describe the *Eckman spiral* (fig. 9.2). Above the level $z = \pi/\gamma$ where $v = 0$ the wind is approximately geostrophic. Below this level the wind direction deviates very considerably from the geostrophic direction; at the surface for instance the deviation is 45°. The quantity π/γ may, therefore, be considered as the approximate depth of the boundary layer. With $f = 7 \times 10^{-5}\,\mathrm{s}^{-1}$ and $K = 10\,\mathrm{m^2\,s^{-1}}$, $\pi/\gamma \simeq 1$ km. Note that in the boundary layer the wind has a component directed generally towards low pressure, a feature which was predicted by the simple argument of §7.3.

Because the approximation of constant K is not a good one, particularly near $z = 0$ (cf. fig. 9.2), the Eckman profile is not accurately followed in

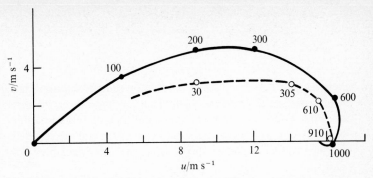

Fig. 9.2. Wind hodograph for the lowest kilometre as measured by Dobson (1914) (dashed line) compared with Eckman spiral (full line). Figures on curves are heights in metres.

practice. Nevertheless as a qualitative description, the solution is a very useful one.

9.4 *The mixing-length hypothesis*

In the last section the eddy transfer coefficient was assumed independent of height. Especially near the surface this assumption is far from true. Some understanding of the reason for this comes from the mixing-length model of turbulence which has been valuable particularly in the near surface layers. In this model it is assumed that an element of fluid at level z moves through an average distance $|l|$ taking its velocity with it. At height $z + l$ it is reabsorbed losing all trace of its original motions. The length l therefore plays somewhat the same part as the mean free path in kinetic theory.

This description implies that

$$u' = -l \frac{\partial \bar{u}}{\partial z}$$

If the turbulence is approximately isotropic $|w'| \simeq |u'|$ and the shearing stress τ will be given by

$$\tau = -\rho \overline{u'w'} = \rho l^2 \left(\frac{\partial \bar{u}}{\partial z} \right)^2 \tag{9.14}$$

In the derivation of (9.14) it must be remembered that l is positive for upward moving parcels of fluid.

In the layer within a few tens of metres of the surface the shearing stress is approximately constant – a layer known as the *constant flux layer*. A further plausible hypothesis is that the size and path of the eddies should be proportional to height above the surface, i.e. $l = \kappa z$ where κ is known as von Karman's constant and has a value of about 0.4. On integrating (9.14) under these assumptions the wind profile is given by

$$\bar{u} = \frac{u_*}{\kappa} \ln\left(\frac{z}{z_0}\right) \tag{9.15}$$

where $u_* = (\tau/\rho)^{1/2}$ is known as the *friction velocity* and the constant of integration z_0 as the *roughness length* since it depends on the surface roughness. Equation (9.15) fits well under conditions of neutral stability. For other situations as might be expected, the wind profile and the associated momentum, heat and water vapour fluxes depend very considerably on the vertical stability.

9.5 Eckman pumping

We return to the Eckman layer and assume for the sake of simplicity that the atmosphere is of uniform density of depth H, and that in the boundary layer of depth d ($d \ll H$) the wind profile is accurately described by (9.12) and (9.13), and that above the boundary layer there is a flow u_g in the x direction, independent of height but varying with the y co-ordinate. Because of friction in the boundary layer, horizontal convergence or divergence occurs which leads through the necessity for continuity to vertical motion.

The continuity equation for a situation where density changes are neglected is (7.19)

$$\frac{\partial w}{\partial z} = -\frac{\partial u}{\partial x} - \frac{\partial v}{\partial y} \tag{9.16}$$

Substituting for u and v from (9.12) and (9.13) and integrating through the boundary layer we have for the vertical velocity w_d at $z = d$

$$w_d = -\int_0^d \frac{\partial u_g}{\partial y} \exp(-\gamma z) \sin \gamma z \, dz \tag{9.17}$$

since $\partial u/\partial x = 0$ and since on a level surface $w = 0$ at $z = 0$.

The vorticity ζ_g of the geostrophic wind above the boundary layer is equal to $-\partial u_g/\partial y$ so that on integration (9.17) becomes

$$w_d = \tfrac{1}{2}\zeta_g\gamma^{-1} \tag{9.18}$$

For typical values (problem 9.6), w_d is a few mm s^{-1}.

The existence of a vertical velocity upwards from the boundary layer has consequences for the flow in the rest of the atmosphere, again because of continuity. Suppose for instance we consider the situation in a region of cyclonic vorticity. There is inflow in the boundary layer towards the centre of the vortex, rising air above the boundary layer and a balancing outflow at higher levels (fig. 9.3). This outflow affects the vorticity ζ_g; the rate of change of ζ_g can be found from (8.36), namely

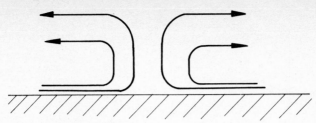

Fig. 9.3. Illustrating Eckman pumping.

$$\frac{\partial \zeta_g}{\partial t} = f \frac{\partial w}{\partial z} \qquad (9.19)$$

where ζ has been neglected compared with f in the r.h.s. of (8.36) (cf. problem 8.16). Integrating (9.19) between the top of the boundary layer $(z = d)$ and the top of the atmosphere $(z = H)$ we have

$$\frac{\partial \zeta_g}{\partial t}(H-d) = -fw_d$$

and on substituting from (9.18)

$$\frac{\partial \zeta_g}{\partial t} = -\frac{f}{2\gamma H}\zeta_g \qquad (9.20)$$

since $d \ll H$.

The result expressed by (9.20) is that the vorticity is reduced with a time constant of $2\gamma H/f$ – the *spin-down time* which is typically several days. The main circulation decays very much more rapidly through this means involving a *secondary circulation* than by other damping mechanisms (problem 9.5). This secondary circulation is driven by friction in the boundary layer, a mechanism known as *Eckman pumping*.

9.6 The spectrum of atmospheric turbulence

The discussion so far in this chapter has been concerned with turbulent motion on the rather small space and time scales associated with the boundary layer. Motion on much larger scales possesses random characteristics and may also be considered as turbulence which therefore occurs in the atmosphere on a very wide range of scales. Fig. 9.4 shows an estimate of the energy spectrum of atmospheric motions on different scales. The minimum in the mesoscale region corresponds to motions in size about 10 km, i.e. about 1 scale height. At smaller scales than this the turbulence is essentially three dimensional and not limited by the atmospheric depth, for larger scales the turbulence is approximately two dimensional. In both these regions simple dimensional arguments have been put forward in an attempt to explain the observed energy spectrum.

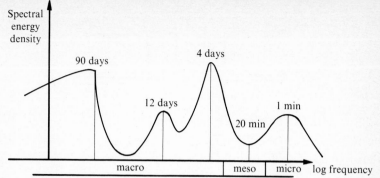

Fig. 9.4. Spectral energy density of atmospheric motions having different time scales. (After Monin, 1973)

First suppose the existence of isotropic homogeneous three dimensional turbulence away from the atmospheric boundary. Further, suppose that in wavenumber space (wavenumber denoted by k), turbulent components at any wavenumber interact with components of similar wavenumber, and that dissipation eventually occurs at very high wavenumbers through viscosity at a rate of ϵ per unit mass. Under these conditions Kolmogoroff assumed that the energy spectrum $E(k)$ depends only on ϵ and simple dimensional considerations require (problem 9.6)

$$E(k) = \alpha_1 \epsilon^{2/3} k^{-5/3} \tag{9.21}$$

where α_1 is a constant. (cf. fig. 9.5).

Fjörtoft (1953) has shown that this energy cascade from lower to higher wavenumbers cannot occur for two dimensional turbulence. In this case it has been postulated that the energy spectrum of turbulence depends only on the rate of transfer of the mean square vorticity $\bar{\zeta}^2$, so that a dimensional argument produces (problem 9.6)

$$E(k) = \alpha_2 (\bar{\zeta}^2)^{2/3} k^{-3} \tag{9.22}$$

where α_2 is a constant. Fig. 9.6 indicates that there is some evidence for this dependence.

General considerations of this kind clearly do not help an understanding of the detail of the atmospheric circulation, but they are of considerable importance in assessing the degree of interaction between motion on different scales in space and time, which assessment has a profound influence on estimates of the fundamental predictability of the general circulation.

Problems

9.1   Show that the product $\rho u w$ is the instantaneous flux of u momentum in the z direction across an element of area parallel to the

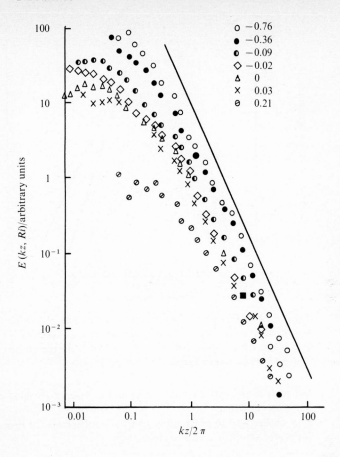

Fig. 9.5. Spectral energy density $E(k)$, of vertical velocity at different wavenumbers k, as measured at heights $z = 1$ m and $z = 4$ m and different values of Richardson number Ri (cf problem 9.7) shown against symbols. The line has a slope of $-5/3$.

earth's surface where ρ, u and w are values appropriate to the position of the element of area. Show that when there is no mean vertical velocity the average momentum flux is $\rho\overline{u'w'}$.

9.2   If T' and m' represent respectively fluctuations in potential temperature and in water vapour mixing ratio, derive the following expressions for the vertical flux of heat Q and water vapour E due to turbulent motion

$$Q = \rho c_p \overline{w'T'}, \qquad E = \rho\overline{w'm'} \qquad (9.23)$$

Near the surface E is also the rate of evaporation. The quantity Q/LE (where L is the latent heat of evaporation) is known as the *Bowen ratio*; it clearly depends on the moisture which is available at

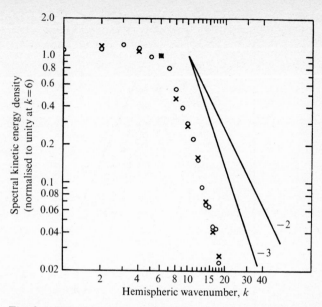

Fig. 9.6. Kinetic energy spectra 500 mb, latitude 50° N for the winter season compared with power laws with exponents -2 and -3 respectively (from Julian *et al.*, 1970). Note the approximate -3 power law dependence at wavenumbers > 6.

the surface and varies from infinity in the desert to a value of about 0.1 over the ocean.

9.3   In a similar way to (9.6), eddy transfer coefficients for heat K_Q and for water vapour K_E can be defined by

$$Q = -\rho K_Q \frac{\partial \theta}{\partial z}, \qquad E = -\rho K_E \frac{\partial m}{\partial z} \qquad (9.24)$$

By considering the physical processes involved, discuss how similar the quantities K (from (9.6)) K_Q and K_E may be expected to be.

9.4   In the constant flux layer the shear stress is ρu_*^2 and the heat flux Q (cf. problem 9.2). Show by dimensional analysis that a length L given by

$$L = c_p \rho T u_*^3 / \kappa g Q \qquad (9.25)$$

can be formed from the quantities where von Karman's constant κ is also included. L is known as the Monin–Oboukhov length. Similarity theory postulates that vertical gradients of heat and momentum should be functions of z/L.

9.5   Put in typical values for ζ_g, K and f in (9.18) and (9.20) and work out values of w_d and of the spin-down time. Show that the spin-down time is very much shorter than the decay time which would occur under damping due to eddy viscosity but without allowing for the possibility of vertical motion and the secondary circulation.

112

9.6   By dimensional analysis deduce (1) the Kolmogoroff $-5/3$ power law equation (9.21), (2) equation (9.22).

9.7   The nature of turbulence near the surface depends on the balance between (1) the energy which can be derived from breakdown of the mean flow and (2) the working of buoyancy forces arising from the transfer of heat away from the surface.

Neglect all horizontal gradients in applying the horizontal momentum equation (9.5) to a region of the boundary layer. Leaving the acceleration and the eddy stress terms only write it as

$$\frac{\partial}{\partial t}(\rho \bar{u}) = -\frac{\partial}{\partial z}\rho \overline{u'w'} \tag{9.26}$$

Multiplying both sides of (9.26) by $\bar{u}$ derive the following expression for the rate of change of kinetic energy

$$\frac{\partial}{\partial t}(\tfrac{1}{2}\rho \bar{u}^2) = -\frac{\partial}{\partial z}(\bar{u}\rho\overline{u'w'}) + \rho\overline{u'w'}\frac{\partial \bar{u}}{\partial z} \tag{9.27}$$

Integrate (9.27) between levels z_1 and z_2 over volume V to give

$$\frac{\partial}{\partial t}\int_V \tfrac{1}{2}\rho \bar{u}^2 \, dV = -[\bar{u}\rho\overline{u'w'}]_{z_1}^{z_2} + \int_V \rho\overline{u'w'}\frac{\partial \bar{u}}{\partial z} dV \tag{9.28}$$

The l.h.s. of (9.28) is the rate of change of kinetic energy in the mean flow. On the r.h.s. the first term is the work done due to the drag at the boundaries z_1 and z_2 and the second term is the rate of transfer of energy from turbulence to the mean flow.

Show therefore by referring to (9.14) that the rate of destruction of turbulence per unit volume by transfer to the mean flow is $-\tau\, \partial \bar{u}/\partial z$.

Now considering the effect of buoyancy forces show that the upward force on an element of unit volume at temperature $\bar{T} + T'$ compared with the surroundings at temperature $\bar{T}$ is $T'g\rho/\bar{T}$. Hence show that the rate of work done on the eddies by buoyancy forces is $Qg/\bar{T}$ per unit volume where $Q = c_p\rho\overline{w'T'}$ (cf. problem 9.2). By comparing these expressions for the rate of work done on the eddies and for the rate of destruction of turbulence show that turbulence will persist if $Ri_F < 1$ where

$$Ri_F = -\frac{g}{\bar{T}}\frac{Q}{c_p\tau\, \partial \bar{u}/\partial z} \tag{9.29}$$

This dimensionless quantity Ri_F is known as the *flux Richardson number*. Now in (9.29) replace τ and Q by the expressions in (9.6) and (9.24) and show that turbulence will persist provided $Ri < K/K_Q$ where

$$Ri = \frac{g\, \partial\theta/\partial z}{\bar{T}(\partial \bar{u}/\partial z)^2} \tag{9.30}$$

Turbulence

The quantity Ri is the *Richardson number*. Under the assumption which is often made that $K \simeq K_Q$ the condition for persistent turbulence is that $Ri < 1$.

10
The general circulation

10.1 Laboratory experiments

In §4.10 (fig. 4.5) we noticed that the atmosphere receives an excess of net radiation in equatorial regions and a deficit in polar regions. A pattern of circulation is, therefore, set up in the atmosphere which transfers heat energy from low latitudes to high latitudes.

Some insight into the type of flow to be expected in the atmosphere of a rotating planet comes from laboratory experiments on the convection which occurs in a rotating fluid. Fig. 10.1 illustrates the dependence of the type of flow on rotation rate for a laboratory experiment in which a water–glycerol solution contained in an annulus heated on the inside and cooled on the outside is rotated at different angular speeds. At low speeds the main flow is axially symmetric and heat is transferred from the centre to the outside mainly by a simple convective cell confined to the boundary layer. As the rotation rate is increased, and Coriolis acceleration becomes more important, waves appear known as *baroclinic waves* rather similar to the Rossby waves discussed in §8.4. These waves at first are regular but at higher rotation rate the preferred wavelength becomes smaller and irregularities in the wave structure occur. The main transport of heat across the annulus takes place within these waves through the mechanism of *sloping convection* (§10.6).

Figure 10.2 illustrates that approximations to these simple regimes occur in the atmosphere. At latitudes near the equator the main transports occur through a meridional circulation; in mid latitudes large scale eddies are the dominant means of transport of heat and momentum.

A further understanding of the mechanisms of energy transfer comes from inspecting various terms in the budget of potential and kinetic energy as described in §3.6 and fig. 3.3. Much of the zonal available potential energy generated by the net source of radiation at the equator and net sink over the poles is converted to eddy available potential energy and eddy kinetic energy by means of the large scale eddy motions at mid latitudes.

In the paragraphs that follow we shall look rather briefly first at a symmetric convective cell, and then at various forms of instability, especially *baroclinic instability*.

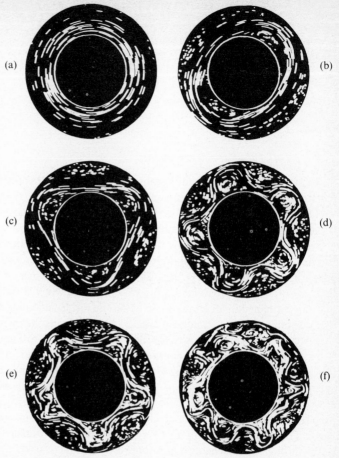

Fig. 10.1. Streak photographs illustrating the dependence of the flow type on rotation rate Ω for a laboratory 'dishpan' experiment. The values of Ω in rad s^{-1} are (a) 0.41; (b) 1.07; (c) 1.21; (d) 3.22; (e) 3.91; (f) 6.4. Working fluid was a water–glycerol solution of mean density 1.037 g cm^{-3} and kinematic viscosity 1.56×10^{-2} cm^2 s^{-1}. The streak photographs show the flow at a depth of 0.5 cm below the free upper surface (see also problem 10.1.) (From Hide & Mason, 1975)

10.2 A symmetric circulation

We want first to find out whether a steady-state solution to the equations of motion can be found for an atmosphere on a rotating planet heated at the equator and cooled at the poles in which the flow shows axial symmetry, i.e. no variation with longitude.

To simplify the discussion (which follows Charney, 1973) we shall assume

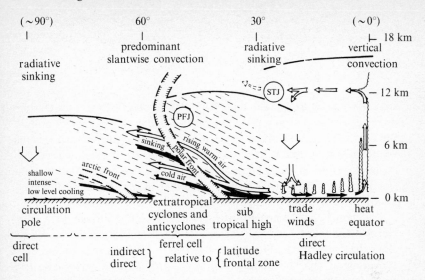

$(\sim 90°)$ 60° 30° $(\sim 0°)$

radiative sinking — predominant slantwise convection — radiative sinking — vertical convection — 18 km

Fig. 10.2. Schematic features of the atmospheric circulation in winter (after Palmén and Newton, 1969) showing Hadley circulation in tropics, sloping convection at mid and high latitudes; showing also polar front associated with which is polar front jet (PFJ) and the sub-tropical jet (STJ) associated with break in tropopause at $\sim 30°$ latitude.

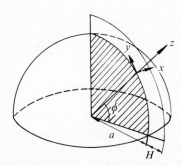

Fig. 10.3. A cross-section of the atmosphere between equator ($\phi = 0$) and pole for the discussion of a symmetric circulation.

the Boussinesq approximation, i.e. the atmosphere will be assumed of uniform density except for buoyancy effects (cf. §8.6). Except near the boundaries the flow will be quasi-geostrophic. Owing therefore, to the temperature gradient with latitude, which we shall assume uniform over the atmospheric depth H, the zonal wind u will show a shear with height given by the thermal wind equation (7.29)

117

The general circulation

$$\frac{\partial u^{(0)}}{\partial z} = -\frac{g}{Tf}\frac{\partial T}{\partial y}$$

$$= -\frac{g}{Tfa}\frac{\partial T}{\partial \phi} \tag{10.1}$$

since $y = a\phi$ (fig. 10.3). The superscript (0) denotes that (10.1) is the zeroth order approximation to u; it will be well obeyed in the main body of the fluid. At the bottom and top boundaries of the atmosphere, however, there are boundary conditions to be satisfied; adjustments to the flow need to be made, therefore, in the Eckman layers near these two boundaries. We follow closely the argument of §9.3 and assume that, as was found in §9.3 for atmospheric conditions, the thickness h of these layers will be $\ll H$.

Taking the upper boundary first, because it is a free surface there cannot be any stress there. Let $\xi = H - z$ denote the distance from the boundary along the z axis. Proceeding in a similar way to §9.3, we seek to find the zonal flow $u_1^{(1)}$ and the meridional flow $v_1^{(1)}$ within the boundary layer (subscript 1 for the *upper* boundary layer) which when added to $u^{(0)}$ will satisfy the conditions

$$\left.\begin{array}{ll}\dfrac{\partial u_1^{(1)}}{\partial \xi} - \dfrac{\partial u^{(0)}}{\partial z} = 0 & (z = H \text{ or } \xi = 0) \\[2mm] \dfrac{\partial v_1^{(1)}}{\partial \xi} = 0 & (\xi = 0)\end{array}\right\} \tag{10.2}$$

$$u_1^{(1)}(\xi) = v_1^{(1)}(\xi) = 0 \quad (\xi \to \infty) \tag{10.3}$$

Equations (10.2) provide for the absence of stress at the top boundary and (10.3) provides for the requirement that the additional flow be confined to the boundary layer.

The equations of motion for the boundary layer flow are (cf. (9.7) and (9.8)

$$\left.\begin{array}{l} K\dfrac{\partial^2 u_1^{(1)}}{\partial \xi^2} + fv_1^{(1)} = 0 \\[3mm] K\dfrac{\partial^2 v_1^{(1)}}{\partial \xi^2} - fu_1^{(1)} = 0 \end{array}\right\} \tag{10.4}$$

where K is the coefficient of eddy viscosity. By a similar method to that in §9.3, the solution which satisfies (10.2) and (10.3) is (problem 10.2)

$$u_1^{(1)}(\xi) = \frac{1}{2\gamma}\left(\frac{\partial u^{(0)}}{\partial z}\right)(\sin\gamma\xi - \cos\gamma\xi)\exp(-\gamma\xi) \tag{10.5}$$

$$v_1^{(1)}(\xi) = \frac{1}{2\gamma}\left(\frac{\partial u^{(0)}}{\partial z}\right)(\sin\gamma\xi + \cos\gamma\xi)\exp(-\gamma\xi) \tag{10.6}$$

118

where $\gamma = \left(\dfrac{f}{2K}\right)^{1/2}$

Equation (10.6) shows that there is poleward flow within the top boundary layer. Because of the variation of γ with latitude, $v_1^{(1)}$ varies with latitude. To compensate for this divergence of the meridional flow, a compensating vertical flow is required. The equation of continuity applied to a thin slice of atmosphere bounded by two circles of latitude is

$$\frac{1}{a\cos\phi}\frac{\partial\,(v_1\cos\phi)}{\partial\phi} + \frac{\partial w}{\partial z} = 0 \tag{10.7}$$

from which, as in §9.5 the vertical velocity w in the upper boundary layer can be found. At the lower surface of this boundary layer it will be given by

$$w(\xi = h) = \frac{1}{a\cos\phi}\frac{\partial}{\partial\phi}\left[\cos\phi\int_0^\infty v_1^{(1)}d\xi\right] \tag{10.8}$$

This flow will, of course, be directed downwards into the fluid (problem 10.4).

We now turn our attention to the lower boundary which is in contact with the planet's surface. The boundary requirements here are that all components of velocity go to zero at the surface and also, because in the steady state there can be no net flow of momentum from the solid surface to the fluid or vice versa, that the average stress in the zonal direction at the surface be zero. This latter condition could, of course, be satisfied if stress at one latitude were balanced by an equal and opposite stress at another latitude. With the symmetry of the situation we are postulating, however, this is not possible and so to satisfy the condition we require the zonal component of the surface stress everywhere to be zero. So far as the stress in the y direction is concerned, provided symmetry is maintained between two hemispheres, stress in one hemisphere is exactly balanced by that in the other.

If we take $u^{(0)} = 0$ at $z = 0$, a flow in the lower boundary layer $u_2^{(1)}(z)$ and $v_2^{(1)}(z)$ (subscript 2 for *lower* boundary) is required which satisfies:

$$\left.\begin{array}{ll} u_2^{(1)}(z) = v_2^{(1)}(z) = 0 & (z = 0) \\[2mm] \dfrac{\partial u_2^{(1)}}{\partial z} + \dfrac{\partial u^{(0)}}{\partial z} = 0 & (z = 0) \end{array}\right\} \tag{10.9}$$

$$u_2^{(1)}(z) = v_2^{(1)}(z) = 0 \quad (z \to \infty) \tag{10.10}$$

The solutions for a lower Eckman boundary layer are in (9.12) and (9.13). Appropriate solutions with the boundary conditions of (10.9) and (10.10) are

$$u_2^{(1)}(z) = -\frac{1}{\gamma}\frac{\partial u^{(0)}}{\partial z}[1 - \cos\gamma z\exp(-\gamma z)] \tag{10.11}$$

119

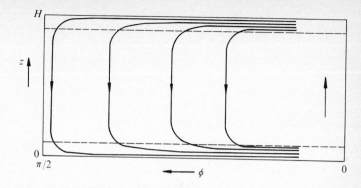

Fig. 10.4. Schematic of symmetric circulation (after Charney, 1973)

$$v_2^{(1)}(z) = -\frac{1}{\gamma} \frac{\partial u^{(0)}}{\partial z} \sin \gamma z \exp (-\gamma z) \qquad (10.12)$$

As before the divergence of $v_2^{(1)}(z)$ implies a vertical flow given by (cf. (10.8)):

$$w(z = h) = -\frac{1}{a \cos \phi} \frac{\partial}{\partial \phi} \left[\cos \phi \int_0^\infty v_2^{(1)}(z) \, dz \right] \qquad (10.13)$$

Notice that $\int_0^\infty v_1^{(1)}(\xi) \, d\xi$ with $v_1^{(1)}(\xi)$ given by (10.6) is exactly equal but opposite in sign to $\int_0^\infty v_2^{(1)}(z) \, dz$ with $v_2^{(1)}(z)$ given by (10.12). The two vertical velocities given respectively by (10.8) and (10.13) are, therefore, equal. Flow in the upper boundary layer is exactly balanced by the reverse flow in the lower boundary layer. Between the two is a gentle sinking motion (fig. 10.4). For a given latitude the various components of u and v are shown in fig. 10.5.

In the above discussion we have followed the flow of fluid and ensured the conservation of mass through the continuity equation. Regarding the flow of heat which we have not considered in detail, our assumption is that the temperature field is maintained mainly by eddy transfer and that, particularly in the main body of the fluid, it will not be significantly disturbed by the slow circulation we have worked out.

The circulation we have found is known as a *Hadley circulation*. As we have already shown in fig. 10.2 such a circulation does occur in the tropical regions of the earth's atmosphere. It is probably also the dominant type of circulation in the lower atmosphere of Venus which is very slowly rotating.

In the simplified circulation which we have described nearly all the transfer of heat and momentum occurs in the boundary layers. Going back to the laboratory experiment, as the rotation rate increases the thickness π/γ of the boundary layers diminishes proportionately. Heat transfer by the fluid

120

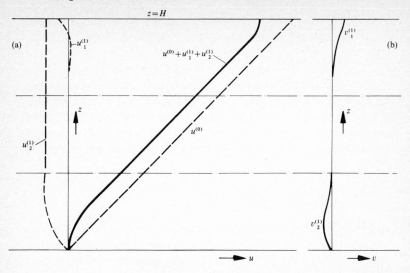

Fig. 10.5. Showing the various components of (a) u and (b) v for the symmetric circulation of Fig. 10.4 (after Charney, 1973). The boundary layers of thickness $\pi/\gamma(\ll H)$ are much exaggerated.

therefore becomes more difficult with increasing rotation rate. We saw from the laboratory experiments of §10.1 that the symmetric circulation becomes unstable as the speed of rotation increases. In the next few sections we shall investigate various forms of instability.

10.3 Inertial instability

For steady zonal flow on a rotating planet in which the geostrophic zonal velocity $\bar{u}(y)$ varies with y but not with z, consider the stability of a 'parcel' displaced with velocity v in the y direction from y_0 to $y_0 + y'$. The equations of motion describing the parcel's change in velocity are (§7.2)

$$\frac{du}{dt} = fv \tag{10.14}$$

and $$\frac{dv}{dt} = f(\bar{u} - u) \tag{10.15}$$

Since $v = dy/dt$, (10.14) may be integrated to give

$$u(y_0 + y') - \bar{u}(y_0) = fy'$$

121

Substituting for $u(y_0 + y')$ in (10.15) we have

$$\frac{dv}{dt} = \frac{d^2 y'}{dt^2} = f\left(\frac{\partial \bar{u}}{\partial y} - f\right) y' \tag{10.16}$$

There will therefore be stable oscillations in y' or exponential growth according as $(\partial \bar{u}/\partial y - f)$ is $<$ or > 0. In the northern hemisphere with positive f the condition for *inertial stability* is therefore that $f - \partial \bar{u}/\partial y > 0$, i.e. that the absolute vorticity of the basic flow be positive. Observations in the atmosphere show that, on the synoptic scale, this is always the case (cf. problem 10.5).

10.4 Barotropic instability

In §10.3, by the simple 'parcel' technique, a particular form of instability was investigated in a similar way to the instability with respect to vertical motion considered in §1.4. There are many other forms of instability which occur in fluid motion. In this section and the next we shall consider the stability of planetary waves which were introduced in §§8.4 and 8.6 in two particular cases. In this section we consider a *barotropic* situation, i.e. one in which there are no horizontal temperature gradients and the basic zonal flow does not vary with height. In the next section we consider a *baroclinic* situation, i.e. one in which horizontal temperature gradients occur with a resulting shear of the zonal wind with height.

For the barotropic case the basic wave equation is the statement that absolute vorticity is conserved, i.e. (8.29):

$$\frac{d(\zeta + f)}{dt} = 0$$

In §8.4 in deriving the basic equation for Rossby waves, we assumed uniform zonal flow $\bar{u}$. Here we allow for a varying zonal flow with latitude so that we may find for what conditions of zonal flow instability may arise.

The perturbation form of (8.29) is

$$\left(\frac{\partial}{\partial t} + \bar{u}\frac{\partial}{\partial x}\right)\zeta' + v'\frac{\partial(\bar{\zeta} + f)}{\partial y} = 0 \tag{10.17}$$

where $\quad \dfrac{d}{dt} = \dfrac{\partial}{\partial t} + \bar{u}\dfrac{\partial}{\partial x}$

Substituting the stream function ψ defined by (8.31)

$$u' = -\frac{\partial \psi}{\partial y}, \qquad v' = \frac{\partial \psi}{\partial x}$$

we have

$$\left(\frac{\partial}{\partial t} + \bar{u}\frac{\partial}{\partial x}\right)\nabla^2 \psi + \frac{\partial \psi}{\partial x}\frac{\partial(\bar{\zeta} + f)}{\partial y} = 0 \tag{10.18}$$

For a wave solution

$$\psi = \mathrm{Re}\{\psi_0(y)\exp[ik(x-ct)]\} \tag{10.19}$$

where the phase velocity c may be complex, i.e.

$$c = c_r + ic_i \tag{10.20}$$

If $c_i > 0$, the wave will grow exponentially with time, i.e. it will be unstable. To find in what situation this may be the case substitute from (10.19) in (10.18) so that

$$(\bar{u}-c)\left(\frac{d^2\psi_0}{dy^2}-k^2\psi_0\right) + \psi_0\frac{\partial(\bar{\xi}+f)}{\partial y} = 0 \tag{10.21}$$

To impose some boundary conditions, suppose that the wave is confined to a zonal channel of width W so that $\psi(y) = 0$ at $y = 0$ and $y = W$. There will now be solutions of (10.21) for certain values of the velocity c. Since the quantity ψ_0 is in general complex let us write it as $\psi_r + i\psi_i$. Multiplying (10.21) by the complex conjugate ψ_0^* of ψ_0, dividing by $\bar{u}-c$, integrating over y and equating real and imaginary parts, we have for the imaginary part

$$\int_0^W\left(\psi_i\frac{d^2\psi_r}{dy^2} - \psi_r\frac{d^2\psi_i}{dy^2}\right)dy = c_i\int_0^W\frac{\partial(\bar{\xi}+f)}{\partial y}\frac{|\psi_0|^2}{|\bar{u}-c|^2}dy \tag{10.22}$$

The integrand on the l.h.s. is

$$\frac{d}{dy}\left(\psi_i\frac{d\psi_r}{dy} - \psi_r\frac{d\psi_i}{dy}\right) \tag{10.23}$$

Since $\psi_i = \psi_r = 0$ at both boundaries the integral of this expression is equal to zero. The integral on the r.h.s., therefore, must be equal to zero. On inspecting the integrand we see that for $c_i > 0$, it is necessary that $\partial(\bar{\xi}+f)/\partial y$ change sign somewhere in the region $0 < y < W$. This therefore is a necessary condition for instability. It was first worked out by Rayleigh, hence it is known as *Rayleigh's criterion*. Since $\partial\bar{\xi}/\partial y = -\partial^2\bar{u}/\partial y^2$ and $\partial f/\partial y = \beta$ (the β-plane approximation introduced in §8.4), $\partial(\bar{\xi}+f)/\partial y$ may be expressed as $\beta - \partial^2\bar{u}/\partial y^2$, so that the condition is that the quantity $\beta - \partial^2\bar{u}/\partial y^2$ should change sign somewhere in the region $0 < y < W$.

It is possible that this barotropic instability is responsible for the formation of tropical depressions in the *intertropical convergence zone* (ITCZ) (fig. 10.6), but it is not the form of instability responsible for the main mid-latitude disturbances.

10.5 Baroclinic instability

In the last section we considered the propagation of planetary waves under conditions of zonal flow which varied with latitude but not with altitude. Following Eady (1949) we now consider the growth of planetary waves under

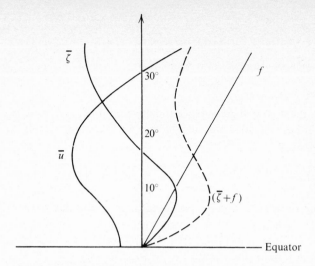

Fig. 10.6. Variation of zonal wind and vorticity near the intertropical convergence zone illustrating the strong shear and the possibility of barotropic instability (after Charney, 1973).

the same general conditions as those considered for the axially symmetric
circulation in §10.2, i.e. in a Boussinesq fluid confined between rigid horizontal planes at $z = -H/2$ and $H/2$ with a horizontal temperature gradient such
that the mean zonal flow $\bar{u}$ possesses a uniform gradient with height (cf.
(7.29), (10.1) and problem 10.6)

$$\frac{\partial \bar{u}}{\partial z} = -\frac{g}{f}\frac{\partial \overline{\ln \theta}}{\partial y} \tag{10.24}$$

where θ is the potential temperature. To further simplify the problem we are
going to neglect the variation of Coriolis parameter with latitude, i.e. we let
$\beta = 0$.

Despite this assumption we shall find that because of the baroclinic situation (i.e. the horizontal temperature gradient) planetary wave solutions exist
and further that under suitable conditions instability occurs such that the
waves will grow. This *baroclinic instability* is the most important form of instability in the atmosphere as it is responsible for the mid-latitude cyclones.

The relevant equations are the vorticity equation (8.45) which with $\beta = 0$
is

$$\left(\frac{\partial}{\partial t} + \bar{u}\frac{\partial}{\partial x}\right)\left(\nabla^2\psi + \frac{f^2}{gB}\frac{\partial^2\psi}{\partial z^2}\right) = 0 \tag{10.25}$$

and the thermodynamic equation

$$\frac{d\ln\theta}{dt} = 0$$

124

which on introducing the perturbations becomes

$$\left(\frac{\partial}{\partial t}+\bar{u}\frac{\partial}{\partial x}\right)(\ln\theta)'+v'\frac{\partial\overline{\ln\theta}}{\partial y}+w'\frac{\partial\overline{\ln\theta}}{\partial z}=0 \tag{10.26}$$

The thermal wind equation (10.24) applied to the perturbation demands

$$\frac{\partial\psi}{\partial z}=\frac{g}{f}(\ln\theta)' \tag{10.27}$$

so that (10.26) becomes

$$\left(\frac{\partial}{\partial t}+\bar{u}\frac{\partial}{\partial x}\right)\frac{f}{g}\frac{\partial\psi}{\partial z}+\frac{\partial\psi}{\partial x}\frac{\partial\overline{\ln\theta}}{\partial y}+w'\frac{\partial\overline{\ln\theta}}{\partial z}=0 \tag{10.28}$$

We look for wave-like solutions

$$\psi=\mathrm{Re}\{\psi_0(z)\exp i(kx+ly-kct)\} \tag{10.29}$$

which on substitution into (10.25) lead to an equation for $\psi_0(z)$, i.e.

$$\frac{f^2}{gB}\frac{\partial^2\psi_0}{\partial z^2}-(k^2+l^2)\psi_0=0 \tag{10.30}$$

which has the solution

$$\psi_0(z)=A\sinh\alpha z+C\cosh\alpha z \tag{10.31}$$

with

$$\alpha^2=\frac{gB}{f^2}(k^2+l^2) \tag{10.32}$$

Because of the rigid boundaries, boundary conditions are imposed $w'=0$ at $z=H/2$ and $-H/2$ which on substitution from (10.31) into (10.28) lead to two simultaneous equations for A and C. These equations are only consistent if the determinant of the coefficients vanishes. An equation for c results namely (problem 10.7)

$$(c-\bar{u}_0)^2=\left(H\frac{\partial\bar{u}}{\partial z}\right)^2\left\{\frac{1}{4}+\frac{1}{H^2\alpha^2}-\frac{\coth H\alpha}{H\alpha}\right\} \tag{10.33}$$

where $\bar{u}_0$ is the value of $\bar{u}$ at $z=0$.

When the r.h.s. of (10.33) is negative, $c-\bar{u}_0$ is wholly imaginary $(=c_i)$. Since the equation for the stream function contains a term $\exp(kc_it)$, under these conditions the wave amplitude will grow exponentially with a time constant $(kc_i)^{-1}$. Note that the real part of c is equal to $\bar{u}_0$ so that the phase velocity of the growing waves is equal to that of the middle of the fluid, i.e. where $z=0$ (sometimes known as the 'steering' level).

The maximum value of kc_i occurs when $l=0$ and $H\alpha=1.61$ (problem 10.8 and fig. 10.7). From (10.32) for $l=0$ the wavelength $\lambda_m=2\pi/k$ of most rapid growth and hence of maximum instability is given by

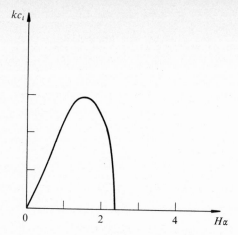

Fig. 10.7. Variation of growth rate kc_i of Eady baroclinic waves as a function of the parameter $H\alpha$ from (10.33).

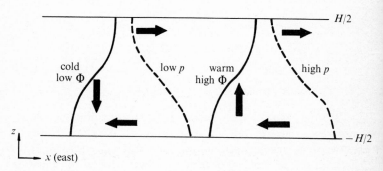

Fig. 10.8. Phases of temperature and pressure in an Eady wave. The motion field is also sketched in. Poleward (northward) movement is associated with rising air and southward movement with falling air.

$$\lambda_m = \frac{2\pi H (gB)^{1/2}}{1.61 f} \qquad (10.34)$$

which for typical atmospheric values (problem 10.9) leads to a wavelength of ~ 4000 km. We should expect disturbances of about this wavelength to be the most common occurring in the atmosphere; a value which fits in well with typical wavelengths for disturbances found in mid latitudes. We further note that mid-latitude cyclones are observed to develop in periods of one to three days which agrees well with typical growth times found from the Eady theory (cf. problem 10.10).

The relative phases and amplitudes of pressure and temperature in the

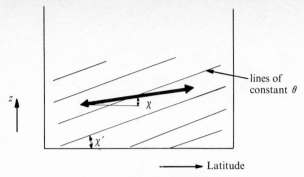

Fig. 10.9. Illustrating sloping convection.

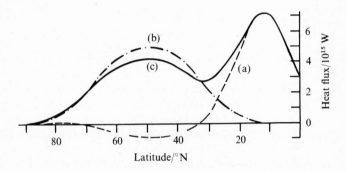

Fig. 10.10. Northward flux of heat in northern hemisphere winter by (a) mean circulations such as the tropical Hadley cell, (b) eddies. Curve (c) is the total flux (after Newton, 1970).

disturbance and also of the vertical and northward velocities are shown in fig. 10.8 (cf. problem 10.11).

10.6 Sloping convection

The baroclinic wave described in the last section is an example of *sloping convection*. Since in the wave, northward moving air is also moving upwards and southward moving air downwards (problem 10.12), the disturbance may be regarded as an eddy in which air parcels are interchanged along a slant path of average slope χ (fig. 10.9). If $\chi < \chi'$ where χ' is the slope of the isentropic surfaces in the undisturbed state, the average potential energy will decrease and kinetic energy will be released into the eddy. If $\chi > \chi'$ eddy kinetic energy will be converted to potential energy. In a rotating system where Coriolis forces dominate over the other inertial or viscous forces, sloping convection is a more effective way of transferring heat energy than axisymmetric motions of the kind considered in §10.2. Transfer of energy by sloping convection

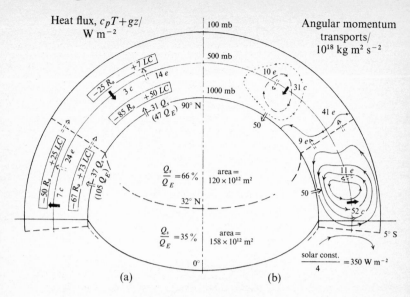

Fig. 10.11. Heat and angular momentum budgets in northern hemisphere winter. In (a), units are averages in W m^{-2} for whole tropical region 5° S – 32° N, and for whole extratropical region. Interior boxes give net atmospheric radiation R_a and condensation heat release LC in layers above and below 500 mb. Dashed arrows, eddy flux e; solid arrows, cell flux c; Q_S, sensible heat flux from earth's surface. For comparison, evaporative flux Q_E is shown in parentheses. (b) shows transports of angular momentum across boundaries, and surface torques. (From Newton, 1970)

dominates over transfer by mean meridional circulations everywhere in the atmosphere outside the Hadley circulation region in the tropics.

10.7 Energy transport

For all seasons there is an excess of net radiation over the tropics and a deficit in polar regions (fig. 4.5). As a result of the atmosphere's circulation heat is transferred from equator to poles. For any given column of atmosphere (including the surface) the energy balance may be divided into a number of terms so that

net radiation = (evaporation − precipitation) × latent heat
+ storage + net transport by atmosphere and ocean.

The net radiation is the difference between incoming solar radiation and outgoing long-wave radiation, as illustrated in fig. 4.5. Storage can be in

128

atmosphere, land or ocean; averaged over one year the net storage will be approximately zero. Transport is either by atmospheric motions or ocean currents. Fig. 10.10 shows the northward flux of heat in the northern winter by meridional motions and by eddy transfer showing that the former are effective in the tropics and the latter at higher latitudes. Heat flux due to ocean currents is difficult to estimate; it is probably less than half that due to the atmosphere. Fig. 10.11 shows various components of the energy budget for different parts of the atmosphere, again in northern winter.

Figure 3.3 shows more detail of the main route of the energy transformations. Zonal available potential energy generated mainly by the gross differences between incoming and outgoing radiation, is converted through the eddy motions to eddy available potential energy which in turn is converted into eddy kinetic energy which is then dissipated.

In the troposphere heat is transported by atmospheric motions down the temperature gradient from the net heat source near the equator to the heat sinks around the poles. In the lower stratosphere the eddy transport of heat is still poleward but reference to fig. 5.1 will show that, because the equatorial stratosphere is colder than the polar stratosphere, this transfer is now against the temperature gradient. This can only happen if the poleward moving air parcels sink and therefore heat adiabatically while the equatorward moving ones rise and therefore cool. Such heat transport is characteristic of a refrigerator rather than a direct heat engine, the stratospheric refrigerator being driven by energy released by the tropospheric heat engine. A diagram similar to fig. 3.3 but for the stratosphere is shown in fig. 10.12, which illustrates energy transformations from the eddy kinetic energy supplied by tropospheric forcing.

10.8 Transport of angular momentum

Apart from small effects due to tidal friction the angular momentum of the earth and its atmosphere remains constant. Since the earth's average angular velocity is very close to being constant, the atmosphere's angular momentum must also be substantially constant (problem 10.14). This means that the average transfer of angular momentum between earth and atmosphere must be close to zero. A poleward transfer of angular momentum must, therefore, occur between the tropical regions where the surface winds are easterly to mid latitudes where they are westerly. In the tropical regions this transfer of angular momentum occurs both through the Hadley circulation and through eddy transfer. In mid latitudes the transfer occurs almost entirely through the large scale eddies (figs. 10.11 & 10.13). The torque on the surface is partially due to friction in the boundary layer and partially due to systematic pressure differences between the western and eastern sides of mountain ranges especially in the northern hemisphere.

For poleward transport of angular momentum in the eddies the quantity

The general circulation

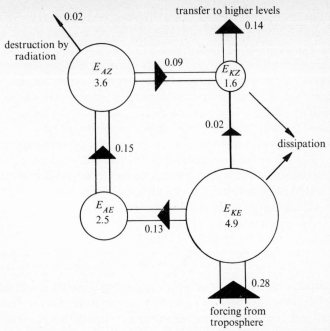

Fig. 10.12. Average storages and main conversions of available potential energy E_A and kinetic energy E_K for 50–10 mb layer in stratosphere as estimated by Crane (1976) from satellite data for period 28 January – 10 February 1973. Units of energy in kJ m^{-2} and of conversion in W m^{-2}. Subscripts Z and E are for zonal and eddy components respectively. Compare with fig. 3.3 for the whole atmosphere.

$\overline{u'v'}$ must be positive, i.e. there must be a significant correlation between positive values of u' and v'. In a symmetrical wave, for instance the fastest growing Eady wave of §10.5 (problem 10.15), this is not the case and the net transport will be zero. If, however, the troughs and ridges in the wave are tilted from south-west to south-east (fig. 10.14) $u' > 0$ when $v' > 0$ and northward transfer of angular momentum will occur.

Notice from fig. 10.13 that much of the zonal momentum transport by eddies occurs against the gradient, thus exhibiting an effective viscosity which is negative. Similar *negative viscosity* situations in which the momentum of eddies is fed into zonal flow have been observed in atmospheres other than that of the earth for instance in the atmospheres of the sun (Starr, 1968), Mars (Leovy, 1969) and Jupiter (Hide, 1974).

Problems

10.1   Consider an annulus rotating at angular velocity Ω containing liquid

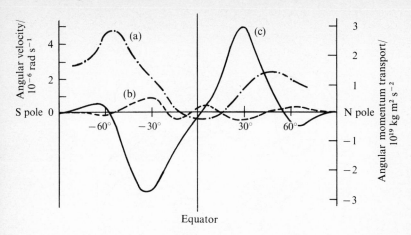

Fig. 10.13(a) Mean angular velocity about polar axis of atmosphere relative to earth's surface; and mean annual northward angular momentum transport by (b) mean circulation, and (c) by eddies. Note the up-gradient transport of angular momentum by eddies. (After Starr, 1968)

as shown in fig. 10.1. The equation of state relating density ρ to temperature T for the liquid is

$$\rho = \rho_0[1 - \alpha(T - T_0)]$$

where α is the thermal expansion coefficient ($\cong 2 \times 10^{-4}\,\mathrm{K}^{-1}$). Derive the thermal wind relationship (cf. §7.6 and problem 7.18) for the annulus in the form

$$\frac{\partial \mathbf{V}}{\partial z} = \frac{\alpha g}{2\Omega}\,\mathbf{k} \wedge \nabla T \tag{10.35}$$

where $\mathbf{V}$ is the fluid's velocity.

If a Rossby number is defined for the annulus (whose interior and exterior radii are a and b respectively) by $Ro = u/\Omega(b-a)$, where u is a typical relative zonal velocity of the fluid, show by substituting in (10.35) that

$$Ro \cong \frac{\alpha g H \Delta T}{2\Omega^2 (b-a)^2}$$

where H is the depth of the fluid and ΔT the difference in temperature across it. Estimate Ro for the situation of fig. 10.1 (a), (c) and (e) respectively for which $\Delta T = 9\,\mathrm{K}$, $b - a = 5\,\mathrm{cm}$, $H = 14\,\mathrm{cm}$ and $\Omega = 0.41$, 1.21 and $3.91\,\mathrm{rad\,s}^{-1}$ respectively.

10.2   Show that (10.5) and (10.6) satisfy (10.4) and the appropriate boundary conditions.

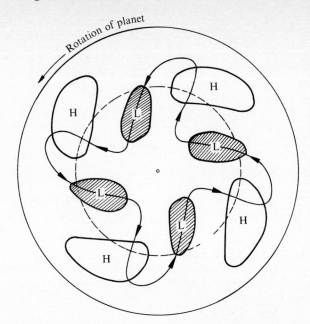

Fig. 10.14. Schematic representation of planetary waves in tropospheric flow. Note the tilting of the waves required to provide a northward momentum flux. Note also the depressions (L) and anticyclones (H) associated with the mean upper tropospheric flow (full line). (After Mintz, 1961)

10.3   Insert typical values into (10.1) to find a typical value of $\partial u^{(0)}/\partial z$. Using this and information from §9.3 compute typical values of $u^{(1)}$ and $v^{(1)}$ from (10.5) and (10.6) and compare with $u^{(0)}$ at the top of the atmosphere.

10.4   Substitute for $v^{(1)}$ in (10.8) from (10.6) and carry out integration to find $w(\xi = h)$. Put in numerical values as in problem 10.3 and estimate a value for $w(\xi = h)$.

10.5   Devise a field of zonal flow for which the absolute vorticity is < 0.

10.6   Write (7.29) in terms of potential temperature and hence deduce (10.24).

10.7   Derive (10.33).

10.8   From (10.33) show that the maximum value of $H\alpha$ for which $c_i \geqslant 0$ is the solution to the equation

$$\frac{x}{2} = \coth \frac{x}{2}$$

[use the identity $\tanh x = (2 \tanh x/2)/(1 + \tanh^2 x/2)$]

10.9   Insert typical values into (10.34) and deduce values for λ_m.

10.10   With values of λ_m as found from problem 10.9, with $H\alpha = 1.61$ and with a typical value of $\partial \bar{u}/\partial z$, from (10.33) find values for $(kc_i)^{-1}$

the time constant of the rate of growth of the disturbances of greatest instability.

10.11 Substitute from (10.31) into (10.28) putting in the boundary condition $w' = 0$ at $z = \pm H/2$. Hence find the ratio C/A and show that for the wave of maximum growth the ratio is imaginary. Hence check the relative phases of quantities in the wave shown in fig. 10.8.

10.12 Show that, in the baroclinic wave of §10.5, rising air is moving northwards and sinking air southwards.

10.13 Air from the equator is moved slowly northward while its angular momentum is conserved. What zonal velocity will it have at latitude 30° N relative to the earth's surface?

10.14 Consider the atmosphere at rest relative to the earth. What is the ratio of its angular momentum to the angular momentum of the solid earth (assume of mean density 5500 kg m^{-3})? Assuming constant zonal velocity for the atmosphere what variation in zonal velocity is required to produce the observed seasonal variation of 2 parts in 10^8 in the earth's rotation rate?

10.15 Show that for the fastest growing Eady wave of §10.5, since $l = 0$, $\overline{u'v'} = 0$, so that the wave is symmetrical and no north–south momentum transport occurs.

11
Numerical modelling

11.1 A barotropic model

In previous chapters the basic equations describing atmospheric structure and motion have been solved for particular situations in order to isolate a number of specific phenomena. In the general case we need to integrate the basic equations with respect to time starting with a given atmospheric situation at a particular time so that a simulation can be provided of the atmospheric behaviour at subsequent times. The task of writing the equations and the boundary conditions in a suitable form and then of solving them with high speed digital computers is known as *numerical modelling*. By comparing the behaviour of the model with that of the real atmosphere the validity of the procedures employed by the model are tested. The most important application of numerical modelling is the development of methods sufficiently reliable and sufficiently fast to be used in routine weather forecasting. The first attempt to build such a model was made by L. F. Richardson in 1918; his book (see bibliography) is still a classic in the field. The first successful forecast was made by Charney, Fjörtoft and von Neumann (1950).

The equation employed for this early forecast was the vorticity equation applied to a barotropic atmosphere, i.e. an atmosphere which is assumed to be homogeneous and of uniform density and in which vertical motion is ignored. Writing the components of the velocity $\mathbf{V}$ in terms of the stream function ψ as defined by (8.31), the vorticity equation (8.29) becomes

$$\frac{\partial}{\partial t}(\nabla^2\psi) + \mathbf{V}.\boldsymbol{\nabla}(\nabla^2\psi + f) = 0 \tag{11.1}$$

Equation (11.1) is a one parameter equation which may be integrated numerically as it stands. It will apply to a level in a real atmosphere if a level can be found where the horizontal divergence of the flow is negligible and where coupling with other levels or with the boundaries may also be neglected. Synoptic flow in the mid troposphere is sufficiently non-divergent that for short periods (11.1) may be employed with some success.

11.2 Baroclinic models

With a barotropic model it is not possible to predict the development of instabilities due to thermal gradients such as were described in §10.5. For these to be present in the model we must describe the motion at more than one atmospheric level, to allow for vertical motion and to use the thermo-dynamic equation.

Under the assumptions that the synoptic scale motions to be described are quasi-horizontal and quasi-geostrophic, a suitable form of the vorticity equation is (8.35) which on using the continuity equation (7.20) and the stream function as defined by (8.31) can be written in isobaric co-ordinates as

$$\frac{\partial}{\partial t}(\nabla^2\psi) + \mathbf{V}.\nabla(\nabla^2\psi + f) - f\frac{\partial\omega}{\partial p} = 0 \qquad (11.2)$$

where in writing the last term ζ has been neglected in comparison with f (problem 8.16).

The thermodynamic equation in isobaric co-ordinates is (using (7.34)):

$$\frac{\partial\ln\theta}{\partial t} + u\frac{\partial\ln\theta}{\partial x} + v\frac{\partial\ln\theta}{\partial y} + \omega\frac{\partial\ln\theta}{\partial p} = \frac{1}{c_p}\frac{dS}{dt} \qquad (11.3)$$

where dS/dt is the rate of increase of entropy due to diabatic processes (§11.6ff).

Now

$$d\ln\theta = \frac{1}{\gamma}d\ln p + d\ln\rho^{-1} \qquad (11.4)$$

from (8.15), and the hydrostatic equation (1.2) when written in terms of the geopotential Φ (problem 7.11) gives

$$\rho^{-1} = -\partial\Phi/\partial p \qquad (11.5)$$

Since the first three terms of (11.3) involve differentiation at constant pressure, with the help of (11.4) and substituting from (11.5), (11.3) becomes

$$\frac{\partial}{\partial t}\left(-\frac{\partial\Phi}{\partial p}\right) + u\frac{\partial}{\partial x}\left(-\frac{\partial\Phi}{\partial p}\right) + v\frac{\partial}{\partial y}\left(-\frac{\partial\Phi}{\partial p}\right) - \frac{\omega B}{g\rho^2} = \frac{1}{\rho c_p}\frac{dS}{dt} \qquad (11.6)$$

where the static stability parameter

$$B = \frac{\partial\ln\theta}{\partial z} = -g\rho\frac{\partial\ln\theta}{\partial p} \qquad (11.7)$$

The geostrophic approximation (7.14) enables the geostrophic stream function ψ defined by (8.31) to be identified with Φ/f (cf. (8.44)) so that (11.6) may be written as

$$\frac{\partial}{\partial t}\left(\frac{\partial\psi}{\partial p}\right) + \mathbf{V}.\nabla\left(\frac{\partial\psi}{\partial p}\right) + \frac{\omega B}{fg\bar{\rho}^2} = -\frac{1}{\bar{\rho}c_p f}\frac{dS}{dt} \qquad (11.8)$$

where in the heating term and the expression for static stability an average density $\bar{\rho}$ has been included.

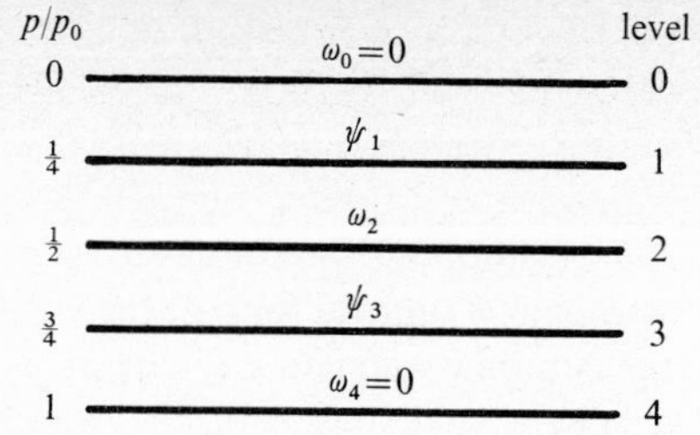

Fig. 11.1. Arrangement of levels and variables for a two parameter baroclinic model.

The vorticity equation (11.2) and the thermodynamic equation (11.8) are the basic equations for numerical integration. To see how this is carried out for a two-level model consider the atmosphere divided up into layers as shown in fig. 11.1 with the levels denoted by subscripts 0 to 4. The equations are solved for stream functions ψ_1 and ψ_3 at the levels 1 and 3. At the top of the atmosphere the boundary condition is $\omega_0 = 0$ and similarly in the absence of orography $\omega_4 = 0$ at the bottom. In terms of finite differences

$$\left(\frac{\partial \omega}{\partial p}\right)_1 \simeq \frac{\omega_2}{\Delta p} \quad \text{and} \quad \left(\frac{\partial \omega}{\partial p}\right)_3 \simeq -\frac{\omega_2}{\Delta p} \tag{11.9}$$

where Δp is a pressure difference of half an atmosphere. Vorticity equations as (11.2) may, therefore, be written for levels 1 and 3, namely

$$\left. \begin{aligned} \frac{\partial}{\partial t} \nabla^2 \psi_1 + \mathbf{V}_1 . \boldsymbol{\nabla}(\nabla^2 \psi_1 + f) - f\omega_2/\Delta p &= 0 \\[2mm] \frac{\partial}{\partial t} \nabla^2 \psi_3 + \mathbf{V}_3 . \boldsymbol{\nabla}(\nabla^2 \psi_3 + f) + f\omega_2/\Delta p &= 0 \end{aligned} \right\} \tag{11.10}$$

where $\mathbf{V}_1 = \mathbf{k} \wedge \boldsymbol{\nabla} \psi_1$ and $\mathbf{V}_3 = \mathbf{k} \wedge \boldsymbol{\nabla} \psi_3$

The finite difference form of (11.8) in terms of ψ_1 and ψ_3 and ω_2 only will be

$$\frac{\partial}{\partial t}(\psi_1 - \psi_3) + \mathbf{V}_2 . \boldsymbol{\nabla}(\psi_1 - \psi_3) - \frac{B\Delta p}{gf\bar{\rho}^2}\omega_2 = \frac{1}{c_p}\frac{\Delta p}{\bar{\rho}f}\frac{dS}{dt} \tag{11.11}$$

where $\mathbf{V}_2 = \mathbf{k} \wedge \boldsymbol{\nabla}\tfrac{1}{2}(\psi_1 + \psi_3)$

If ω_2 is eliminated between (11.10) and (11.11), two equations for ψ_1 and ψ_3 result which may be integrated numerically.

11.3 Primitive equation models

The baroclinic model described in §11.2 although less restrictive than the barotropic model of §11.1 still involves a number of approximations. In

136

particular the motion is assumed to be quasi-geostrophic. This restriction can be removed by writing the equations in a more basic form, namely: the horizontal momentum equations (7.8)

$$\frac{\partial \mathbf{V}}{\partial t} + \mathbf{V}.\nabla V + \omega \frac{\partial \mathbf{V}}{\partial p} + f\mathbf{k} \wedge \mathbf{V} = -\nabla \Phi + \mathbf{F} \tag{11.12}$$

the continuity equation (7.20)

$$\nabla.\mathbf{V} + \frac{\partial \omega}{\partial p} = 0 \tag{11.13}$$

the hydrostatic equation (1.2)

$$\frac{\partial \Phi}{\partial p} + \frac{1}{\rho} = 0 \tag{11.14}$$

and the thermodynamic equation (11.6)

$$\frac{\partial}{\partial t}\left(-\frac{\partial \Phi}{\partial p}\right) + \mathbf{V}.\nabla\left(-\frac{\partial \Phi}{\partial p}\right) - \frac{B\omega}{g\rho^2} = \frac{1}{c_p\rho}\frac{dS}{dt} \tag{11.15}$$

In these equations $\mathbf{V}$ is the horizontal wind and ∇ refers to differentiation at constant pressure.

The five equations (11.12) to (11.15) ((11.12) is of course two equations) involve five unknowns, two components of $\mathbf{V}$, ω, Φ and ρ.

Although ordinary sound waves have been eliminated from the solutions of this set of equations by the use of the hydrostatic equation (11.14) and the neglect of vertical accelerations, solutions remain for a wide range of different motions including for instance the acoustic and the gravity waves described in §8.3. These solutions are eliminated for the simplified models of §11.1 and §11.2 through the quasi-geostrophic approximation.

With primitive equation models care has to be taken that the solutions are not swamped by spurious gravity or acoustic waves which may arise from errors in the initial data or from computational instability. To overcome these problems it is necessary to introduce an initial motion field which satisfies (11.12); actual observations of wind must not be employed as they stand. Further a time step has to be chosen for the integration process which is short enough to avoid computational instability for the fast-moving gravity or acoustic waves. For this reason time steps need to be considerably shorter than with quasi-geostrophic models with similar horizontal resolution. Primitive equation models, therefore, demand much more computer time than comparable quasi-geostrophic ones.

In (11.12) a frictional term $\mathbf{F}$ has been included and in (11.15) the diabatic heating term. A number of physical processes are involved in these terms. Further an adequate description of the boundary conditions at the surface is required. Some indications of how these matters may be dealt with in

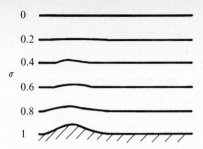

Fig. 11.2. The σ co-ordinate system for a five level model.

a numerical model are presented in the sections which follow. More detail may be found in the works listed in the bibliography.

11.4 Inclusion of orography

It is necessary in a model to take into account the varying height of the surface. In the isobaric co-ordinate system where pressure is the vertical co-ordinate, the varying surface pressure due to the varying height of the surface is not easy to handle. For numerical models, therefore, a co-ordinate σ is employed, defined as

$$\sigma(x,y,t) = \frac{p(x,y,t)}{p_s(x,y,t)} \tag{11.16}$$

where p_s is the surface pressure. This system is known as the *sigma co-ordinate system* (fig. 11.2, problem 11.1). σ is a non-dimensional parameter; the lower boundary is always at $\sigma = 1$ where the vertical σ velocity $\dot\sigma = 0$.

11.5 Convection

As integrations in the model proceed situations occur where the vertical temperature gradient is super adiabatic. These are especially likely in regions where there is strong heating of the surface. To make allowance for the effect of convection, *convective adjustment* is carried out in the models such that the temperature profile is adjusted to the dry adiabatic (§1.4) under conditions of low water vapour content, and to the saturated adiabatic (§3.2) when the relative humidity is equal to or close to 100%. This adjustment has to be made, of course, in such a way that the total energy is unchanged, i.e. the sum of the total potential energy and latent heat released by the adjustment must be zero.

11.6 Moist processes

Water vapour evaporated from the surface and moved about by atmospheric

138

motions plays a very significant part in the atmosphere's energy budget because of (1) the release of latent heat on condensation and (2) the formation of clouds which alter the radiation budget (cf. chapter 6). A numerical model, therefore, needs to keep track of the water content in vapour and liquid form. An equation rather similar to the continuity equation may be written for the water vapour mixing ratio m

$$\frac{\partial}{\partial t}(\rho m) + \mathbf{V}.\nabla(\rho m) + \omega \frac{\partial(\rho m)}{\partial p} = -C + E + D \qquad (11.17)$$

where C is the rate of condensation, E the rate of evaporation into the volume from the surface, if the surface is contained within the volume in question and D is a diffusion term (cf. §11.8). The rate of condensation C is determined by comparing the water vapour mixing ratio m with the saturation mixing ratio m_s appropriate to the atmospheric temperature. If $m > m_s$, the excess water is assumed to condense and precipitate to the surface. Associated with the rate of condensation C there is a latent heat release LC, where L is the latent heat of condensation, which needs to be included in the term dS/dt in the energy equation (11.15).

11.7 Radiation transfer

The radiative processes which contribute to atmospheric heating and which need to be included in the diabatic term in (11.15) divide into two parts (§2.1) namely the absorption of solar radiation and the exchange of terrestrial radiation. Simple examples of calculations for both of these were given in chapter 4.

Absorption of solar radiation in a clear atmosphere is mainly by ozone, water vapour and carbon dioxide. Detailed calculations of the amount of absorption by an atmospheric path for any of these gases may be made using the methods outlined in chapter 4 and the data in appendices 8, 9 and 10. For incorporation into numerical models, however, these methods are too expensive in computer time and further simplification is necessary. The various bands are fairly well separated from each other so that for the sun at zenith angle θ the solar energy absorbed in an atmospheric path is

$$\cos \theta \{f_1[u(O_3)] + f_2[u^*(H_2O)] + f_3[u^*(CO_2)]\} \qquad (11.18)$$

where f_1, f_2 and f_3 are empirical functions. The path length

$$u = \sec \theta \int_{\text{path}} c\rho \, dz \qquad (11.19)$$

(c being the absorber concentration at height z where the density is ρ) is employed in the calculation of ozone absorption without any correction (cf. appendix 9). For water and carbon dioxide, however, allowance needs to be made for the dependence of absorption on the pressure p (§4.4); the simplest

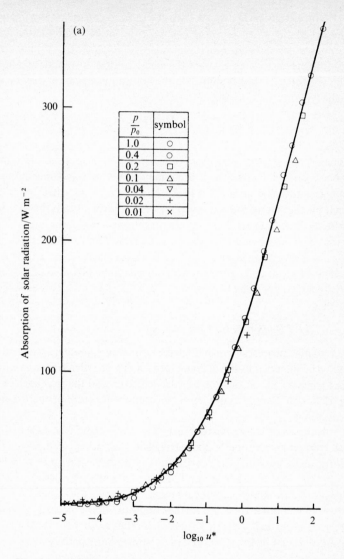

Fig. 11.3. (a) The absorption of solar radiation by an atmosphere containing water vapour as a function of modified path length u^* (11.20) in $\mathrm{g\,cm^{-2}}$ wave clouds, 67 , for a range of values of p/p_0, showing that the pressure-scaling employed is a good approximation.

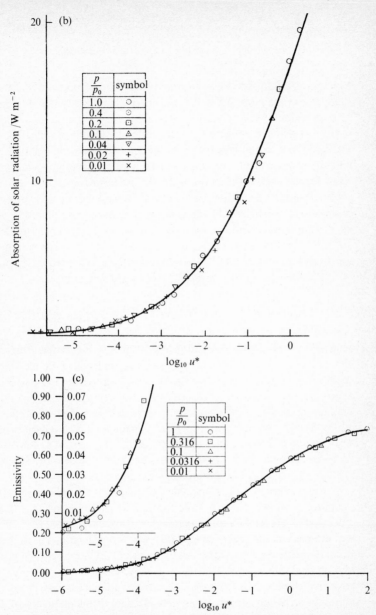

(b) The same as (a) but for an atmosphere containing carbon dioxide with modified path length u^* (in g cm^{-2}) and with $x = 0.68$. (c) The emissivity of a slab of atmosphere at 220 K containing water vapour showing the usefulness of the modified path length u^* ($x = 0.7$) in g cm^{-2} as a parameter. The wavenumber range 550–800 cm^{-1} has been omitted since in that range CO_2 absorption is dominant. (All after Manabe and Wetherald, 1967)

way of doing this is to employ a modified path length

$$u^* = \sec \theta \int_{\mathrm{path}} c\rho \left(\frac{p}{p_0}\right)^x dz \quad (0.5 < x < 1) \tag{11.20}$$

which is scaled so as to be appropriate to a standard pressure p_0. A suitable value of x in the range $0.5 < x < 1$ is chosen empirically (fig. 11.3).

Absorption of solar radiation not only occurs in the direct beam but also in the diffuse beam reflected or transmitted by clouds or reflected from the surface. The effect of absorption of the upward travelling radiation is calculated by employing (11.18) for the total path and including a factor to allow for the average direction of the diffuse beam.

The transfer calculation for terrestrial radiation also needs to be enormously simplified. In chapter 4 band models to assist in the integration over frequency were described. Even these are, however, too time consuming for normal use with large numerical models. A simpler approach is to divide the spectrum into a small number of spectral intervals denoted by $\Delta\nu_i$ and following (4.6) to write for the upward or downward flux at level z

$$F(z) = \sum_i \left\{ \int_z^{z_0} \pi B_i(z') \, d\tau_i[u^*(z',z_0)] + \pi B_i(z_0)\tau_i[u^*(z,z_0)] \right\} \tag{11.21}$$

where $B_i(z')$ is the Planck function appropriate to the temperature at level z' integrated over $\Delta\nu_i$, u^* is the pressure-scaled path length as defined by (11.20) (with $\sec \theta = 5/3$, cf. §4.3) for the absorbing gas (either CO_2, H_2O or O_3) appropriate to the spectral interval $\Delta\nu_i$. τ is an empirical transmission function determined by experiment and appropriate to a reasonable range of atmospheric conditions (fig. 11.3) and z_0 denotes the boundary which may be the ground or the surface of a cloud. From the upward and downward fluxes so calculated, the radiative contribution to the heating term dS/dt in (11.15) may be found by differencing the fluxes as in (2.4).

It will be clear from the discussion of chapter 4 that (11.21) involves a lot of approximation both in its basic formulations (see for instance Rodgers, 1967) and in its use of a single parameter u^* to describe the absorbing path for each gas and spectral interval. However, for calculations in the lower atmosphere the accuracy achieved by using (11.21) even with only one or two spectral intervals is satisfactory, especially considering the much larger errors which may arise from the difficulty of adequately specifying the atmospheric parameters in the models. Because of the difficulty of describing the condensation and precipitation process, water vapour content cannot be specified with much precision and cloud cover even less precisely. The best that can be done at the moment regarding cloud cover is to relate the percentage of cloud cover to the humidity in such a way that the overall average cloud cover agrees with the average from observations.

It is also necessary to assume very simple radiative properties for the clouds, namely a value of albedo depending only on the actual height and an infrared emissivity of unity for all clouds.

11.8 Sub grid scale processes

We saw in §9.6 that there is interaction in the atmosphere between motions on a wide range of scales. With a numerical model, therefore, which inevitably possesses a finite grid size it is necessary to make allowance for the effect of motions on smaller scales than the grid size. The terms $\mathbf{F}$ in (11.12) and D in (11.17) were included to allow for such transfer of momentum and water vapour by the small scale eddies; also there will be a contribution to the term dS/dt in (11.15) from the eddy transfer of heat. The simplest formulation of both horizontal and vertical eddy transfer employs eddy diffusion coefficients K which when multiplied by the appropriate gradients give the transfer, e.g. (9.6). Typical values of K for the troposphere for sub grid scale parameterization lie in the range $1-10\,\mathrm{m^2\,s^{-1}}$. The Ks can, of course, be allowed to vary with height, the static stability, or with other quantities describing the local flow. In particular, K for momentum can be formulated in a way which preserves the k^{-3} spectral distribution of turbulent energy (§9.6).

11.9 Transfer across the surface

Momentum, heat and water vapour are transferred across the earth's surface; equations for the transfer of each are required.

First some method is required to arrive at the surface temperature. Since the effective heat capacity of a land surface is small, its temperature T_s is determined by its local heat balance, i.e.

$$(1 - A)I_S + F^{\downarrow} - \sigma T_s^4 + Q_s - LE = 0 \tag{11.22}$$

where σ is the Stefan–Boltzmann constant, A is the surface albedo, I_S the incident solar radiation (after allowing for absorption or reflection in the atmosphere or the clouds above as described in §11.7), $F^{\downarrow}$ the downward flux of terrestrial radiation from the atmosphere or from clouds calculated as described in §11.7, Q_s the vertical transfer of heat from the atmosphere to the surface by eddy processes, E the rate of evaporation of water vapour from the surface and L the appropriate latent heat. The emissivity of the surface for the terrestrial radiation is assumed to be unity. Methods of calculating Q_s and E are described below. For the ocean surface, by contrast, an infinite heat capacity is assumed and the temperature of the sea surface is assumed to be constant with time; values for it are included in the data initially given to the model.

If the model's lowest level is close to the surface and well within the boundary layer, say at 100 m altitude or so, the flux of momentum across the surface or the surface stress τ will be (following Manabe, 1969 and Holloway

& Manabe, 1971)

$$\tau = -\rho c_D |V|V \tag{11.23}$$

where ρ is the density, c_D the drag coefficient (see problem 11.3), V the vector wind appropriate to the first level. Similar expressions may be written for the flux of heat Q_s and water vapour E, namely

$$Q_s = c_p \rho c_D |V| \Delta\theta \tag{11.24}$$

and

$$E = \rho c_D |V| \Delta m \tag{11.25}$$

where $\Delta\theta$ and Δm are respectively the difference in potential temperature and water vapour mixing ratio between the surface (assumed saturated at surface temperature) and the first level. Equations (11.23) to (11.25) are applicable to neutral conditions; the effect of different stabilities can be taken into account by allowing c_D to vary with stability. Equation (11.25) also only applies to a wet surface. This is, of course, all right over the ocean. Over the land allowance must be made for variation in the soil moisture content W (mass per unit area) which can be found from the equation

$$\frac{\partial W}{\partial t} = P - E \tag{11.26}$$

where P is the precipitation rate determined from the moist processes of §11.6 and E the evaporation rate. If W exceeds a given value the excess water is assumed to run off. If W exceeds a certain lower value W_k the surface is assumed wet; if $W < W_k$ the evaporation rate is assumed to be reduced by the ratio W/W_k from the value given by (11.25). Snow cover can be allowed for in a similar way to soil moisture.

11.10 Other models

The models described so far in this chapter when applied to the whole atmosphere involve a very large number of grid points and so for integrations for only a short time, require a great deal of computer time. To avoid this problem some success has been achieved with *two dimensional models*, i.e. models in which atmospheric parameters are averaged around latitude circles so that only variations with latitude and height are considered. The main difficulty with such models is that of devising means for dealing with the transport of heat and momentum by the large scale eddies which have to be related in simple ways to zonally averaged quantities. Two dimensional models are particularly useful in upper atmospheric modelling in which changes in composition due to complex photochemical reactions (cf. §5.5) have to be included and for which integrations over long periods (e.g. several years) are required (see e.g. Harwood and Pyle, 1975).

Another possibility particularly applicable to the upper atmosphere is to employ a *spectral model* in which horizontal variations are represented in terms of a series of spherical harmonics. Because for the upper atmosphere most of the variation occurs in the first few terms of such a series, i.e. those describing the first few wavenumbers (the wavenumber refers to the number of complete cycles of a variation which occur around a latitude circle), a substantial amount of computing time can be saved by using such transformations.

Problems

11.1 Show that the following equation will transform from the isobaric vertical co-ordinate system to the sigma system

$$\mathbf{\nabla}_p = \mathbf{\nabla}_\sigma - \frac{\sigma}{p_s} \mathbf{\nabla} p_s \frac{\partial}{\partial \sigma} \qquad (11.27)$$

where $\mathbf{\nabla}$ refers to horizontal differentiation only. Hence derive the horizontal momentum equation

$$\frac{d\mathbf{V}}{dt} + f\mathbf{k} \wedge \mathbf{V} = -\mathbf{\nabla}\Phi + \frac{\sigma}{p_s} \mathbf{\nabla} p_s \frac{\partial\Phi}{\partial\sigma} \qquad (11.28)$$

and the continuity equation

$$\mathbf{\nabla} \cdot (p_s \mathbf{V}) + p_s \frac{\partial\dot\sigma}{\partial\sigma} + \frac{\partial p_s}{\partial t} = 0 \qquad (11.29)$$

where $\dot\sigma = \dfrac{d\sigma}{dt}$

11.2 Consider a vertical column of unit cross-section above a location where the surface pressure is p_s. By considering the convergence of mass into the column show that

$$\frac{\partial p_s}{\partial t} = -\int_0^1 \mathbf{\nabla} \cdot (p_s \mathbf{V}) \, d\sigma$$

Show that the same result may be obtained by integrating the continuity equation (11.29).

11.3 Show from (9.15) that under neutral conditions c_D in (11.23) is given by

$$c_D = \left[\frac{\kappa}{\ln(z/z_0)} \right]^2$$

Hence find its value if $z = 100$ m, $z_0 = 1$ cm and $\kappa = 0.4$.

12
Global observation

12.1 What observations are required?

To provide adequate and accurate initial conditions for numerical models and to enable comparison to be made between the models and the real atmosphere, detailed knowledge of the atmospheric state over the whole atmosphere is required continuously in time. We need to know the three dimensional fields of motion, of density and of composition. Clearly, measurement in complete detail is not possible. Further, some parameters are related to each other. For instance, away from the equatorial regions, the geostrophic approximation is a good one so that motion need not necessarily be measured independently if the pressure field is specified. Also, if the field of temperature is known together with the pressure at some reference surface, the density and pressure can be deduced at all levels from the hydrostatic equation (1.4) (problem 12.1). The following specification of what is required is considered realistic from the measurement point of view; it will also go a long way towards satisfying the requirements of the numerical models being prepared for the first Global Atmospheric Research Programme global experiment (chapter 13).

Atmospheric state parameters	*Accuracy (RMS error)*
Wind components	$\pm 3 \, \mathrm{m\,s^{-1}}$
Temperature	$\pm 1 \, \mathrm{K}$
Pressure of reference level	$\pm 0.3\%$
Water vapour pressure	$\pm 1 \, \mathrm{mb}$
Sea surface temperature	$\pm 0.2 \, \mathrm{K}$

Measurements of these parameters (or their deduction from other observations) are required every 12 hours, with at least one measurement every 100 km in the horizontal and at least eight data levels in the vertical (surface, 900, 700, 500, 200, 100, 50 and 20 mb). Information is also required on precipitation, cloud cover, surface conditions (e.g. snow or ice cover) and elements of the radiation budget.

12.2 Conventional observations

Over the land areas of the world much of the information about the atmosphere comes from a high density of surface observations of pressure, temperature, humidity, wind and cloud cover together with measurements twice per day from a large number of *radiosondes*. These are balloon-borne packages containing simple instruments for the measurement of pressure, temperature and humidity, together with a radio transmitter and a radar reflector so that the package can be tracked to give details of the wind. Measurements up to ~ 30 km (~ 10 mb) are possible with the radiosonde. A much more sparse network of rocket-sondes – sondes released from high altitude (up to ~ 90 km) rockets which then descend on parachutes – enables some observations to be made at higher levels.

More specialized measurements, especially of atmospheric composition, are possible from larger balloons, aircraft or rockets. Spectroscopic observations, particularly of absorbed sunlight, from the surface or from various altitudes also provide information regarding composition. The study of meteor trails and other techniques has enabled density and wind in the 70–100 km region to be deduced.

12.3 Remote sounding from satellites

The great advantage of a satellite as a measurement platform is that good coverage in space and time can be obtained. From a *geostationary* satellite, orbiting at 35 000 km altitude so that it remains directly above a fixed point on the equator, continuous observation of about a quarter of the atmosphere is possible. A *polar orbiting* satellite such as Nimbus or Tiros in a circular orbit at about 1000 km altitude makes about fourteen orbits a day and can view all parts of the atmosphere at least twice per day.

Radiation from the earth-atmosphere system reaches an orbiting satellite over a wide range of wavelengths. In the ultraviolet, visible and near infrared, solar radiation is scattered and reflected from the surface, from clouds, from aerosol (particles suspended in the atmosphere) and from molecules. In the infrared and microwave regions, at wavelengths almost completely separated from those where solar radiation is present (fig. 2.1), radiation is emitted again from the surface, clouds and molecules. Over this wide range of wavelengths a great deal of information is contained about the structure and composition of the atmosphere below. Interpretation of these *remote sounding* observations is often complex and difficult, but as we have seen, they possess the enormous advantage compared with conventional observations that a satellite can cover a very large area in a short time.

The first weather satellite was launched in 1960; it carried television cameras for viewing clouds. For the first time complete pictures of the cloud associated with large weather systems were seen. Such information is now produced routinely from a large number of satellites. Detailed cloud pictures of Mars, Venus and Jupiter from space probes have also provided a surprisingly large amount of information about the circulation of their atmospheres.

Infrared imaging systems on satellites enable the emitted radiation field to

Fig. 12.1. Image of the earth in the infrared ($10-12\,\mu$m) taken by NOAA satellite in geostationary orbit on 5 July 1975. Warm surfaces are dark and cold surfaces bright. Notice the hurricane in the centre of the picture.

be mapped in various spectral intervals. In atmospheric *window* regions, for instance between 10 and 12 μm in wavelength, the radiance received corresponds quite closely to the Planck black-body function at the surface or cloud-top temperature (fig. 12.1). Suitable small corrections have to be made for atmospheric transparency and surface emissivity (problem 12.2). Measurements over broad spectral regions provide information abou the earth's radiation budget over different areas (e.g. fig. 4.4).

Measurements at higher spectral resolution in different regions give more precise information about the earth's temperature structure and composition. These we shall consider in turn in the following sections.

12.4 Remote sounding of atmospheric temperature

At any frequency in the infrared where an atmospheric constituent possesses

Global observation

(so as to not allow Transmission)

strong absorption the radiation intensity leaving the top of the atmosphere is a function of the distribution of the emitting gas and the distribution of temperature through the atmosphere.

Constituents such as carbon dioxide with strong absorption bands at $15\,\mu m$ and $4.3\,\mu m$ and molecular oxygen with absorption near 5 mm wavelength in the microwave region, are very nearly uniformly mixed, at least up to levels ~ 90 km altitude. Provided that local thermodynamic equilibrium (LTE) applies (§5.6) the emitted intensity in these bands can be considered to be dependent only on the atmospheric temperature distribution.

At a given wavenumber $\tilde{\nu}$ the intensity of radiation $I_{\tilde{\nu}}$ (known as the *radiance*) received by a satellite-mounted radiometer viewing vertically downwards is given by (4.4)

$$I_{\tilde{\nu}} = \int_0^\infty B_{\tilde{\nu}}(T) \frac{d\tau_{\tilde{\nu}}(z, \infty)}{dz}\, dz + B_{\tilde{\nu}}(T_s)\tau_{\tilde{\nu}}(0, \infty) \qquad (12.1)$$

where T_s is the surface temperature.

The variable $y = -\ln p$, where p is the pressure in atmospheres, is a more convenient height-dependent variable to use, so that (12.1) can be written

$$I_{\tilde{\nu}} = \int_0^\infty B_{\tilde{\nu}}(T)\kappa(y)\, dy + B_{\tilde{\nu}}(T_s)\tau_{\tilde{\nu}}(0, \infty) \qquad (12.2)$$

$\kappa(y)$ is known as the *weighting function*. For a spectral region with a uniform absorption coefficient $k_{\tilde{\nu}}$ independent of altitude, (problem 12.3 and fig. 12.2).

$$\kappa(y) = k_{\tilde{\nu}}cpg^{-1} \exp\left(-k_{\tilde{\nu}}cpg^{-1}\right) \qquad (12.3)$$

→ y dependence!
not accurately known !

where c is the mass mixing ratio of the absorbing constituent. The function $\kappa(y)$ has a peak at $k_{\tilde{\nu}}cpg^{-1} = 1$, i.e. where the optical depth measured from the top of the atmosphere is unity.

Inspection of fig. 12.3 illustrates how remote temperature sounding is achieved. Moving from the atmospheric window near $800\,\text{cm}^{-1}$ $(12.5\,\mu m)$ to smaller wavenumbers, the mean absorption coefficient of carbon dioxide gradually increases so that the average level being monitored moves up in altitude until the most absorbing region is reached – the Q branch near 667 cm^{-1} – where the radiation largely originates in the stratosphere. For figs. 12.3(a) and 12.3(b), the temperature decreases with altitude to the tropopause, then increases again, while for fig. 12.3(c) the temperature increases with altitude all the way up. In fig. 12.4 the application of results to the remote sounding of tropospheric temperature is illustrated. Notice also that the spectra of fig. 12.3 contain information from which the water vapour and ozone distribution may be inferred. Similar observations have also been made of the atmosphere of Mars for which, being almost pure carbon dioxide, measurements in the $15\,\mu m$ carbon dioxide band are particularly appropriate.

It is important that radiance measurements be made with adequate

149

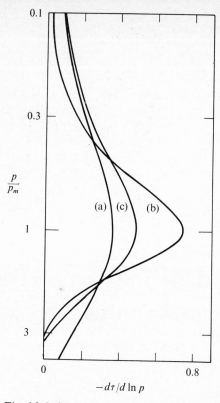

Fig. 12.2. 'Weighting functions' appropriate to a radiometer sounding the atmosphere by observing radiation emitted vertically upwards from the atmosphere for (a) an atmosphere with uniform absorption coefficient, (12.3), (b) a frequency in the wing of a pressure broadened spectral line, problem 12.4, (c) an Elsasser band, problem 12.5. Note the greater vertical resolution of (b). p_m is the pressure at which the functions peak.

accuracy. For the 15 μm region, to obtain temperatures to 1 K, radiance must be measured with a maximum error of considerably less than 1% (problem 12.6). This implies careful radiometric calibration and also means, for conventional instrumentation, that the spectral resolving power cannot be very high. It will be noticed in fig. 12.3 that individual rotational lines cannot be resolved so that much information, particularly about the high atmosphere, is being lost. Houghton & Smith (1970) have overcome this problem by employing a less conventional spectroscopic technique – selective chopping with absorbing gas cells – achieving a spectral resolution approximately equal to that of the absorption lines themselves. Fig. 12.5 shows the weighting functions appropriate to the selective chopper radiometer on Nimbus 4; fig. 12.6

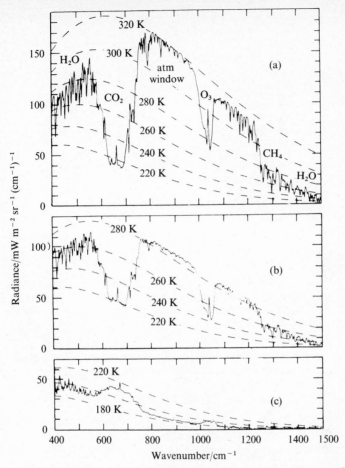

Fig. 12.3. Thermal emission from the earth plus atmosphere emitted vertically upwards and measured by the infrared interferometer spectrometer on Nimbus 4, (a) over Sahara, (b) over Mediterranean, (c) over Antarctic. The radiances of black bodies at various temperatures are superimposed. (From Hanel *et al.*, 1971)

and figs. 8.5 and 8.6 show results from that instrument illustrating the global coverage obtained for an altitude region of the atmosphere largely inaccessible to conventional observation. A further development of the technique known as the pressure modulator radiometer enables remote sounding temperature measurements to be made to ~ 90 km altitude.

The major problem in the interpretation of radiance measurements in infrared regions is the very variable presence of clouds. Considerable effort has gone into developing techniques for retrieving temperature profile information when broken cloud is present. Because non-precipitating clouds are

151

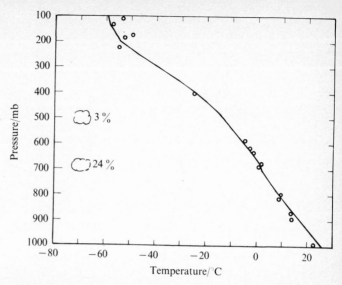

Fig. 12.4. Temperature profile (full line) retrieved from measurements of radiance in the 4.3 μm and 15 μm CO_2 bands by the high resolution infrared sounder and from measurements in the 5 mm O_2 band by the scanning microwave spectrometer on Nimbus 6.

The sounding is for 48° N, 15.2° E on 23 August 1975; the circles are a nearby radiosonde. The presence of 3% of high cloud and 24% of medium cloud was deduced from the retrieval process. (After Smith, 1976)

largely transparent at millimetre wavelengths, measurements of oxygen emission at 5 mm wavelength do not suffer from this drawback, and are, therefore, particularly important for monitoring the troposphere (cf. fig. 12.4).

12.5 Remote measurements of composition

Various absorption bands of water vapour are available for observation in the infrared and microwave regions; given temperature information from carbon dioxide or oxygen emission, some details of the water vapour distribution may be inferred by similar techniques. The ozone distribution may also be studied in this way. Information regarding the minor constituents such as CH_4, N_2O, CO, NO, NO_2, SO_2, which are important in studies of ozone photochemistry (§5.5) or pollution studies, is much more hidden. For stratospheric levels observation from satellites of emission from the atmosphere's limb is a promising technique (fig. 12.7 and problems 12.9 and 12.10); the problem of the very variable emitting background of the earth's surface or clouds is thereby eliminated.

152

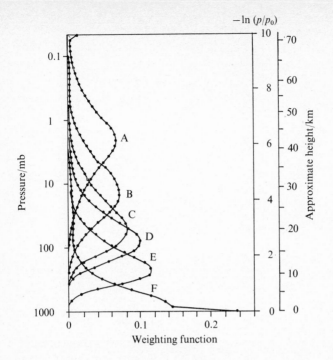

Fig. 12.5. Weighting functions for the six channels of the selective chopper radiometer on Nimbus 4. (After Barnett, 1974)

For remote sounding observations of ozone, particularly at high levels, the ultraviolet region is important. Solar radiation back-scattered from the atmosphere is strongly affected by ozone absorption in the 200 to 300 nm region. To illustrate the method consider a simplified situation in which an ultraviolet spectrometer is observing the radiation leaving the atmosphere in a vertical direction, the sun being overhead (fig. 12.8). At the levels considered attenuation of solar radiation by ozone absorption will be much greater than that due to scattering (problem 12.11); further, single scattering only will be considered.

At a given frequency ν the intensity of solar radiation at the top of the atmosphere is $I_{S\nu}(0)$ and at a level where the pressure is p will be

$$I_{S\nu}(p) = I_{S\nu}(0) \exp\{-k_\nu n(p)\} \qquad (12.4)$$

k_ν being the absorption coefficient per molecule of ozone and $n(p)$ being the number of molecules above the level of pressure p.

From the molecules in a layer of thickness dp at level p the radiation scattered vertically upwards will be proportional to $I_{S\nu}(p)$ and to dp, say it is $a I_{S\nu}(p)\, dp$ (fig. 12.8). Further attenuation will occur on traversing the return path to the satellite, so that neglecting any scattering from the lower

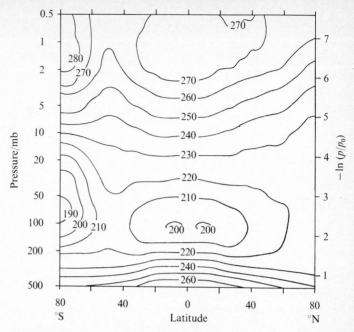

Fig. 12.6. Meridional cross-section of temperature from 500–0.5 mb in altitude and 80° S to 80° N in latitude as inferred from measurements by the selective chopper radiometer on 21 September 1971. Although at an equinox period notice the substantial differences in stratospheric temperature between the hemispheres. (From Barnett, 1974)

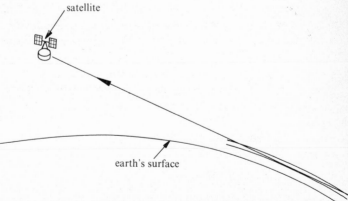

Fig. 12.7. Illustrating limb sounding of the earth's atmosphere. Measurements of emission from the atmosphere's limb have the advantages of (1) a very long emitting path is viewed so that constituents present in very small concentrations can be studied, (2) near zero radiation background beyond the limb.

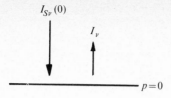

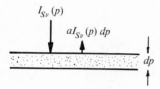

Fig. 12.8

atmosphere, the measured intensity will be

$$I_\nu = I_{S\nu}(0)a\int_0^\infty \exp\{-2k_\nu n(p)\}\,dp \tag{12.5}$$

For purposes of illustration consider a uniform mixing ratio of ozone, i.e.
$n(p) = n'p$ say, n' being a constant and

$$I_\nu = I_{S\nu}(0)a\int_0^\infty p\exp(-2k_\nu n'p)\,d(\ln p) \tag{12.6}$$

The quantity under the integral is very similar to the weighting function of
(12.3); it has a peak at $p = (2k_\nu n')^{-1}$. By choosing a set of wavelengths for
which k_ν is different, it therefore becomes possible to monitor the ozone concentration at different levels. Such measurements have been made from the
back-scatter ultraviolet spectrometer mounted on Nimbus 4 (Krueger, Heath
& Mateer, 1973).

12.6 Observations from remote platforms

To deduce the atmospheric density field, in addition to global measurements
of temperature, the pressure at a reference surface is required. Conventional
surface pressure observations are only very sparse over the oceans. To provide
global coverage of pressure measurements on a reference surface a fleet of
floating buoys or of constant density balloons (i.e. balloons of fixed volume
which float at a given density level) with appropriate instrumentation has
been proposed; these will be interrogated by orbiting satellites which act as a
communication link to these and other observation platforms in remote areas.

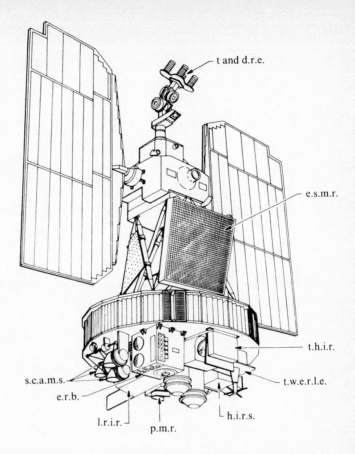

Fig. 12.9. The Nimbus 6 satellite showing the location of the experiments: s.c.a.m.s. (scanning microwave spectrometer); e.r.b. (earth radiation budget experiment); l.r.i.r. (limb radiance infrared radiometer); p.m.r. (pressure modulator radiometer); h.i.r.s. (high resolution infrared sounder); t.w.e.r.l.e. (tropical wind energy conversion and reference level experiment); t.h.i.r. (temperature and humidity infrared radiometer); e.s.m.r. (electrically scanning microwave radiometer); and t and d.r.e. (tracking and data relay experiment). The circular ring on which the experiments are mounted is approximately 1 m in diameter (from NASA, Nimbus 6 User's Guide).

Because of the breakdown of the geostrophic approximation in tropical regions, good measurements of the wind field are required, at least at low latitudes. Observation of the motion of suitable clouds provides some information, other information can be provided from the motion of free floating balloons.

Figure 12.9 shows instrumentation for remote sounding and for

interrogation of remote platforms mounted on the Nimbus 6 satellite
launched in 1975. For complete specification of the atmospheric state, co-
incident observations need to be made in a large number of spectral regions.
The potential of such satellite instrumentation is enormous so far as observation of the whole of the earth's atmosphere and planetary atmospheres is
concerned.

Problems

12.1 The density near the 100 mb level is deduced from surface pressure
measurement combined with temperature measurements throughout
the rest of the atmosphere. What is the percentage error in density
resulting from (a) a 1 K error in temperature at 100 mb, (b) a 1 K
error in temperature at all levels below 100 mb, (c) a 3% error in
pressure measurement at the surface?

12.2 From the information contained in problem 4.11, and assuming the
water vapour mixing ratio at any altitude is proportional to p^3 where
p is the atmospheric pressure in atm, calculate the transmission of a
vertical column of atmosphere in the 11 μm window.

If $B(p)$ is the black-body function at 11 μm appropriate to the
temperature at the level where the pressure is p atm, for an atmospheric lapse rate, $B(p)$ is approximately equal to $B_0 p$ where B_0 is the
black-body function at the surface temperature. Using this expression and (4.3) calculate the intensity leaving the top of the atmosphere at 11 μm. Hence estimate the error in the surface temperature
deduced from a satellite observation if atmospheric absorption is
ignored.

12.3 Obtain the expression in (12.3) for the 'weighting function' for an
atmosphere with uniform absorption coefficient $k_{\bar{\nu}}$.

12.4 Show that in the wing of a collision broadened line of strength s,
and width γ, the absorption coefficient is given by (cf. (4.7))

$$k_{\bar{\nu}} = \frac{s\gamma}{\pi(\nu - \nu_0)^2}$$

Hence show that the weighting function $\kappa(y)$ appropriate to such a
frequency is of the form

$$\kappa(y) = 2\left(\frac{p}{p_m}\right)^2 \exp\left[-\left(\frac{p}{p_m}\right)^2\right]$$

where $p_m^2 = \dfrac{g\pi p_0(\nu - \nu_0)^2}{s\gamma_0 c}$

(quantities as defined in (4.9) and (12.3)).

12.5 A useful model of a set of absorption lines is that due to Elsasser
which assumes uniform line strength and spacing. For this model for
collision broadened lines the transmission of a path between the

157

level at pressure p and the top of the atmosphere is given by

$$\tau = 1 - \frac{2}{\pi^{1/2}} \int_0^{\beta p} \exp(-x^2)\, dx$$

where β depends on line strengths, widths, spacing and absorber concentration. Show that, for this case, the weighting function is given by

$$\kappa(y) = \left(\frac{2}{\pi}\right)^{1/2} \frac{p}{p_m} \exp\left(-\frac{p^2}{2p_m^2}\right)$$

12.6 Calculate from the Planck function for a source at 240 K the percentage accuracy in radiance measurements which is required to achieve a 1 K accuracy in temperature for wavelengths of (a) 4.3 μm, (b) 15 μm, (c) 5 mm.

12.7 A radiometer observing the atmosphere possesses an optical system with a collecting area A, field of view Ω steradians, spectral bandwidth $\Delta \tilde{\nu}$ centred at wavenumber $\tilde{\nu}$, and mean transmission τ_0. Incident radiation from a black body of radiance $B_{\tilde{\nu}}$ is chopped sinusoidally. Show that the signal:noise ratio for measurement in t seconds is

$$\frac{S}{N} = \frac{B_{\tilde{\nu}} A\Omega\Delta\tilde{\nu}\tau_0\, t^{1/2}}{2^{3/2}(\text{NEP})}$$

where NEP is the noise equivalent power of the detector (i.e. the r.m.s. radiation power incident on the detector which gives a signal equal to the noise in a 1 Hz bandwidth).

12.8 The noise equivalent temperature (NET) of a radiometer is the change of source temperature which causes a change in signal just equal to the noise. Calculate for a source temperature of 240 K the NET for an instrument with the following characteristics:

$$A = 100\,\text{cm}^2, \qquad \Omega = 10^{-4}\,\text{sr}, \qquad \tau_0 = 0.5,$$

$$\Delta\tilde{\nu} = 1\,\text{cm}^{-1}\ \text{at}\ 667\,\text{cm}^{-1}, \qquad \text{NEP} = 10^{-10}\,\text{W},$$

$$t = 4\,\text{s}.$$

12.9 A satellite mounted radiometer is observing the limb of the atmosphere. Show that the path of atmosphere traversed through the limb (i.e. $\int \rho\, dx$ where x is along the line of sight) is approximately 70 times that in a vertical path above the tangent point of the path.

12.10 Show that for the path of problem 12.9 the mean pressure along the path as defined by (4.17) is $2^{-1/2}p_0$ where p_0 is the highest pressure along the path. What is the horizontal distance between points where the pressure $2^{-1/2}p_0$?

12.11 From the formula in problem 6.14, for a path near 60 km altitude and 0.27 μm wavelength compare attenuation due to Rayleigh scattering with that due to ozone absorption (ozone concentration at

60 km, 10^{10} cm^{-3}, ozone molecular absorption cross-section 10^{-17} cm^2, other information in the appendices).

12.12　In an atmosphere with spherical symmetry where density ρ and hence refractive index n varies with distance r from the centre of the sphere, for a ray passing through the atmosphere the quantity nd is a constant where d is the length of the perpendicular from the centre of the sphere to the tangent to the path at the point where the refractive index is n. From this expression determine the actual position of the lowest point of a ray passing through the atmosphere to have (a) 10 km, (b) 20 km, as its lowest point.

12.13　A spacecraft emitting radio signals is occulted by the planet Venus. By studying the time taken for reception of the signals it is possible to trace the paths of rays through the Venus atmosphere during occultation and so to deduce the variation of density with altitude. Find the highest pressure on the Venus atmosphere from which such rays will emergy. (Refractive index of CO_2 at STP for radio frequencies $= 1.00049$.)

13
Atmospheric predictability and climatic change

13.1 Short-term predictability

Given an adequate description of the initial conditions, numerical models as described in chapter 11 can satisfactorily predict the atmospheric state for periods ahead of hours up to a few days. For longer periods deficiencies in the models become apparent. The most fundamental of these is the parameterization of sub grid scale motions. Even when, as larger computers become available, the grid size is reduced, there will always be smaller scale motion to be considered which, because of the lack of a comprehensive theory describing the interactions between different scales which always occur in turbulent motion, can only be dealt with in a crude empirical manner.

The existence of this fundamental difficulty raises the question as to how predictable in principle is the atmosphere's general circulation. To attack this problem, the Global Atmospheric Research Programme (GARP) has been formulated, the purpose of which is, over a period of at least a year, to observe the atmosphere as completely as possible so that the performance of

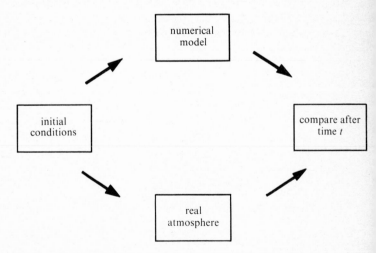

Fig. 13.1. The global experiment of the Global Atmospheric Research Programme (GARP).

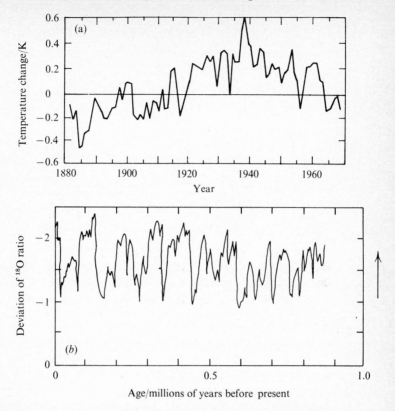

Fig. 13.2. Trends in global climate. (a) Changes of annual mean temperature for northern hemisphere for the past century. The data is mainly due to Budyko (after National Academy of Sciences, 1975). (b) Fluctuations in global ice volume during the last million years as recorded by changes in isotopic composition of fossil plankton in deep sea core. The ordinate is deviation in parts per thousand of $^{18}O : ^{16}O$ ratio from a standard value. Larger negative values of the ratio is interpreted as meaning smaller global ice volume. Notice the rapid deglaciations (after Shackleton and Opdyke (1973)).

numerical models can be tested (fig. 13.1) and light thrown upon the fundamental question of predictability.

13.2 Longer-term variations

Substantial changes in the atmosphere's circulation have taken place in historic and prehistoric times (fig. 13.2). The large number of possible causes for such changes are illustrated in fig. 13.3. These causes can be extra-terrestrial, i.e. changes in solar radiation, or they can arise from couplings within the

161

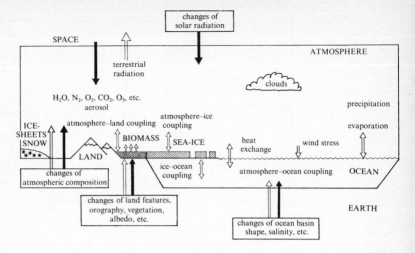

Fig. 13.3. Schematic illustration of the components of the coupled atmosphere–ocean–ice–earth climatic system (from GARP Publication Series No. 16 (see Bibliography)).

atmosphere itself or between the atmosphere, the ocean, the ice or the land surface. Because of the long periods associated with the deep ocean circulation or changes in the physical or chemical properties of the surface, the characteristic times associated with different possible mechanisms can be anything from a few years to periods comparable with the age of the earth itself.

Of particular interest are climatic changes which might occur as a result of activity through altering atmospheric composition (e.g. the increase in carbon dioxide due to burning of fossil fuels) the properties of the land surface or the run-off from rivers. Without any good understanding at the present time of the causes for natural variations, it is clearly difficult to assess the possible effects of such artificial alterations.

13.3 Atmospheric feedback processes

Processes exist in the atmosphere possessing characteristics of both positive and negative feedback. As an example of positive feedback consider a land mass which is covered by ice or snow. Because of the resulting high albedo of the surface much of the incident solar radiation is reflected back out of the atmosphere, thus (in the absence of any other mechanisms) reducing the surface temperature further leading to possible further increases in ice or snow cover. For an example of negative feedback consider an increase in solar radiation incident on the earth's surface which leads to a higher surface temperature and hence increased evaporation. Resulting from the higher water content of the atmosphere is an increase in cloudiness which in turn reduces the solar radiation reaching the surface.

162

Because of the existence of these complicated feedbacks in the atmosphere the questions have been raised as to how stable is the atmosphere's average state and whether there are other quasi-stable states between which the atmosphere may fluctuate. Inspection of the record of climatic history suggests that such may in fact be the case.

13.4 Climate modelling

General circulation models such as those described in chapter 11, even if they are going to predict the general climatic state over periods of a few months or a few years, must not only model the atmosphere but must also include the ocean's circulation and the interaction between ocean and atmosphere. If good confidence in such models is established by experiments such as those planned in the context of the Global Atmospheric Research Programme it will be possible to employ these to predict the effect of various natural events (e.g. volcanic eruptions) or artificial changes (e.g. deforestation of a given region).

Because of the enormous amount of computer time which such general circulation models consume, for the investigation of longer-term climatic change, it will be necessary to employ simpler models which in some part of their formulation may need to be more empirical, but which in any case can be tested over short periods against the more complete general circulation models. By such means it is hoped that eventually elucidation of the principal causes of climatic change will be possible so enabling us to understand much more fully our atmospheric environment on which so much of the quality of human life depends.

Appendices

1. SOME USEFUL PHYSICAL CONSTANTS AND DATA ON DRY A

Avogadro's number	$6.022 \times 10^{23}\,\text{mol}^{-1}$
Loschmidt number	$2.687 \times 10^{25}\,\text{m}^{-3}$
Boltzmann constant k	$1.381 \times 10^{-23}\,\text{J}\,\text{K}^{-1}$
Planck constant h	$6.6262 \times 10^{-34}\,\text{J}\,\text{s}$
Gas constant R	$8.3143\,\text{J}\,\text{K}^{-1}\,\text{mol}^{-1}$
Stefan–Boltzmann constant σ	$5.670 \times 10^{-8}\,\text{J}\,\text{m}^{-2}\,\text{K}^{-4}\,\text{s}^{-1}$
Velocity of light c	$2.998 \times 10^{8}\,\text{m}\,\text{s}^{-1}$
Ice point	$273.15\,\text{K}$
Earth's mean radius	$6371\,\text{km}$
Mean solar angular diameter	31.99 minutes of arc
Standard surface gravity	$9.806\,65\,\text{m}\,\text{s}^{-2}$
Standard pressure p_0	$1013.25\,\text{mb}\ (\equiv 1.013\,25 \times 10^{5}\,\text{Pa})$

Data on dry air

Apparent molecular weight	28.964
Gas constant for dry air	$287.05\,\text{J}\,\text{kg}^{-1}\,\text{K}^{-1}$
Specific heats of dry air:	
at constant pressure c_p	$1005\,\text{J}\,\text{kg}^{-1}\,\text{K}^{-1}$
at constant volume c_v	$718\,\text{J}\,\text{kg}^{-1}\,\text{K}^{-1}$
Ratio of specific heats γ	1.40
Density of dry air at 273 K and 1013 mb pressure	$1.293\,\text{kg}\,\text{m}^{-3}$
Viscosity (at STP)	$1.73 \times 10^{-5}\,\text{kg}\,\text{m}^{-1}\,\text{s}^{-1}$
Kinematic viscosity (at STP)	$1.34 \times 10^{-5}\,\text{m}^{2}\,\text{s}^{-1}$
Thermal conductivity (at STP)	$2.40 \times 10^{-2}\,\text{W}\,\text{m}^{-1}\,\text{K}^{-1}$

Refractive index n of dry air at 1013 mb, 273 K, and wavelength of 1 μm
$$= 1 + 289.2 \times 10^{-6}$$

At other wavelengths $\lambda\,\mu$m Eldén's formula may be used

$$\{n(\lambda) - 1\} \times 10^{6} = 64.328 + 29\,498.1\,(146 - \lambda^{-2})^{-1}$$
$$+ 255.4(41 - \lambda^{-2})^{-1}$$

To obtain n at other temperatures and pressures note that $n - 1$ is proportional to density. For air containing water vapour there is a correction which is usually negligible at visible and infrared wavelengths but which becomes significant in the mm wavelength region. (For more information see Penndorf (1957).)

2. PROPERTIES OF WATER VAPOUR

Molecular weight	18.015
Latent heat of fusion at 273 K	$3.34 \times 10^5 \, \mathrm{J \, kg^{-1}}$
Latent heat of vaporization at 273 K	$2.500 \times 10^6 \, \mathrm{J \, kg^{-1}}$
Specific heat of liquid water at 273 K	$4.218 \times 10^3 \, \mathrm{J \, kg^{-1} \, K^{-1}}$
Specific heat of ice at 273 K	$2.106 \times 10^3 \, \mathrm{J \, kg^{-1} \, K^{-1}}$
Density of ice at 273 K	$917 \, \mathrm{kg \, m^{-3}}$

Table A2.

After *Smithsonian Meteorological Tables* Smithsonian Institute, Washington D.C. 1958

Saturation vapour pressure (mb) *over pure liquid water*

°C	−30	−20	−10	0	+10	+20	+30	+40
0	0.5088	1.2540	2.8627	6.1078	12.272	23.373	42.430	73.777
+1	0.5589	1.3664	3.0971	6.5662	13.119	24.861	44.927	77.802
+2	0.6134	1.4877	3.3484	7.0547	14.017	26.430	47.551	82.015
+3	0.6727	1.6186	3.6177	7.5753	14.969	28.086	50.307	86.423
+4	0.7371	1.7597	3.9061	8.1294	15.977	29.831	53.200	91.034
+5	0.8070	1.9118	4.2148	8.7192	17.044	31.671	56.236	95.855
+6	0.8827	2.0755	4.5451	9.3465	18.173	33.608	59.422	100.89
+7	0.9649	2.2515	4.8981	10.013	19.367	35.649	62.762	106.16
+8	1.0538	2.4409	5.2753	10.722	20.630	37.796	66.264	111.66
+9	1.1500	2.6443	5.6780	11.474	21.964	40.055	69.934	117.40

Saturation vapour pressure (mb) *over pure ice*

°C	−100	−90	−80	−70	−60
0	1.403×10^{-5}	9.672×10^{-5}	5.472×10^{-4}	2.615×10^{-3}	1.080×10^{-2}
+1	1.719×10^{-5}	1.160×10^{-4}	6.444×10^{-4}	3.032×10^{-3}	1.236×10^{-2}
+2	2.101×10^{-5}	1.388×10^{-4}	7.577×10^{-4}	3.511×10^{-3}	1.413×10^{-2}
+3	2.561×10^{-5}	1.658×10^{-4}	8.894×10^{-4}	4.060×10^{-3}	1.612×10^{-2}
+4	3.117×10^{-5}	1.977×10^{-4}	1.042×10^{-3}	4.688×10^{-3}	1.838×10^{-2}
+5	3.784×10^{-5}	2.353×10^{-4}	1.220×10^{-3}	5.406×10^{-3}	2.092×10^{-2}
+6	4.584×10^{-5}	2.796×10^{-4}	1.425×10^{-3}	6.225×10^{-3}	2.380×10^{-2}
+7	5.542×10^{-5}	3.316×10^{-4}	1.662×10^{-3}	7.159×10^{-3}	2.703×10^{-2}
+8	6.685×10^{-5}	3.925×10^{-4}	1.936×10^{-3}	8.223×10^{-3}	3.067×10^{-2}
+9	8.049×10^{-5}	4.638×10^{-4}	2.252×10^{-3}	9.432×10^{-3}	3.476×10^{-2}

(− 50°C to 0°C, see over page.)

Appendices

Table A2. (continued)

°C	− 50	− 40	− 30	− 20	− 10	0
0	3.935×10^{-2}	1.283×10^{-1}	3.798×10^{-1}	1.032	2.597	6.107
+ 1	4.449×10^{-2}	1.436×10^{-1}	4.213×10^{-1}	1.135	2.837	
+ 2	5.026×10^{-2}	1.606×10^{-1}	4.669×10^{-1}	1.248	3.097	
+ 3	5.671×10^{-2}	1.794×10^{-1}	5.170×10^{-1}	1.371	3.379	
+ 4	6.393×10^{-2}	2.002×10^{-1}	5.720×10^{-1}	1.506	3.685	
+ 5	7.198×10^{-2}	2.233×10^{-1}	6.323×10^{-1}	1.652	4.015	
+ 6	8.097×10^{-2}	2.488×10^{-1}	6.985×10^{-1}	1.811	4.372	
+ 7	9.098×10^{-2}	2.769×10^{-1}	7.709×10^{-1}	1.984	4.757	
+ 8	1.021×10^{-1}	3.079×10^{-1}	8.502×10^{-1}	2.172	5.173	
+ 9	1.145×10^{-1}	3.421×10^{-1}	9.370×10^{-1}	2.376	5.623	

3. ATMOSPHERIC COMPOSITION

Table A3. *Normal composition of clean dry air near sea level*

Gas	Volume mixing ratio
Nitrogen (N_2)	0.780 83
Oxygen (O_2)	0.209 47
Argon (Ar)	0.009 34
Carbon dioxide (CO_2)	0.000 33
Neon (Ne)	18.2×10^{-6}
Helium (He)	5.2×10^{-6}
Krypton (Kr)	1.1×10^{-6}
Xenon (Xe)	0.1×10^{-6}
Hydrogen (H_2)	0.5×10^{-6}
Methane (CH_4)	2×10^{-6}
Nitrous oxide (N_2O)	0.3×10^{-6}
Carbon monoxide (CO)	0.1×10^{-6}

For ozone (O_3) distribution see §5.5.

Water vapour (H_2O)

The above table is for dry air. Average water vapour distribution is shown in fig. A3.1. The water vapour mass mixing ratio of the stratosphere is much more uniform than in the troposphere and is in the range 2 to 5×10^{-6}.

4. RELATION OF GEOPOTENTIAL TO GEOMETRIC HEIGHT

	Geopotential metres (gpm)										
Latitude	10 000 m	20 000 m	30 000 m	40 000 m	50 000 m	100 000 m	200 000 m	300 000 m	400 000 m	500 000 m	600 000 m
0°	10 036	20 104	30 204	40 336	50 500	101 811	206 948	315 577	427 874	544 029	664 243
30°	10 023	20 077	30 163	40 282	50 432	101 672	206 656	315 115	427 225	543 174	663 161
45°	10 009	20 050	30 123	40 228	50 465	101 534	206 363	314 653	426 576	542 318	662 080
60°	9 996	20 024	30 083	40 174	50 297	101 395	206 071	314 191	425 927	541 465	661 000
90°	9 983	19 997	30 043	40 120	50 229	101 256	205 779	313 730	425 280	540 613	659 923

(From *American Institute of Physics Handbook*, McGraw-Hill 1972)

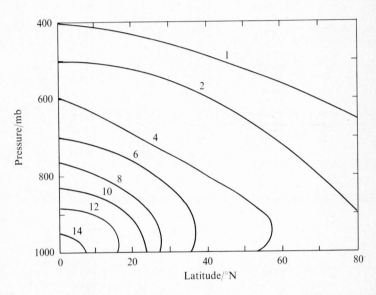

Fig. A3.1. Contours of average water vapour mass mixing ratio in $g\,kg^{-1}$ (after Newell *et al.*, 1972).

5 MODEL ATMOSPHERES (0–105 km)

In the following tables geopotential height, pressure and temperature are plotted for different latitudes and months for typical northern hemisphere conditions. The three quantities are related by the hydrostatic equation (1.4).

Sources of data are Oort & Rasmusson (1971), Labitzke *et al.* (1972), and COSPAR (1972).

For an indication of the difference between the hemispheres see fig. 12.6. The tabulated data of appendix 5 are plotted in fig. 5.1(a) and (b).

Appendices

March

Pressure (mb)	Geopotential height (km) 10° N	Temp. (K)	Geopotential height (km) 40° N	Temp. (K)	Geopotential height (km) 70° N	Te (K)
1000	0.097	299.1	0.127	280.5	0.108	253.2
950.0	0.545	296.6	0.550	278.2	0.503	253.9
900.0	1.016	294.1	0.992	275.8	0.909	254.6
850.0	1.510	291.5	1.454	273.6	1.333	254.8
700.0	3.152	283.1	2.996	266.4	2.770	250.0
500.0	5.864	267.4	5.546	250.5	5.171	236.1
400.0	7.579	256.5	7.146	238.8	6.680	226.7
300.0	9.679	241.4	9.102	225.2	8.549	218.9
200.0	12.41	219.3	11.70	216.4	11.12	218.3
100.0	16.56	196.2	16.10	215.4	15.60	219.5
70.0	18.65	200.9	18.35	215.7	17.96	219.9
50.0	20.62	205.5	20.47	216.1	20.18	220.4
30.0	23.75	217.0	23.70	218.9	23.21	215.6
10.0	30.90	235.0	30.62	226.0	30.27	226.0
7.0	33.38	243.0	33.00	232.0	32.66	229.0
5.0	35.79	250.0	35.27	237.0	34.91	231.0
3.0	39.60	261.0	38.92	248.0	38.39	235.0
1.0	48.29	274.0	47.27	266.0	46.20	249.0
0.70	51.13	271.0	50.06	267.0	48.81	253 0
0.50	53.79	267.0	52.66	264.0	51.31	254.0
0.30	57.67	252.0	56.53	255.0	55.12	252.0
0.10	65.23	222.0	64.42	234.0	62.90	232.0
0.07	67.52	214.0	66.81	228.0	65.30	226.0
0.05	69.64	208.0	69.07	223.0	67.50	222.0
0.03	72.66	203.0	72.32	217.0	70.78	216.0
0.01	79.13	196.0	79.13	203.0	77.56	208.0
0.007	81.15	196.0	81.19	200.0	79.72	206.0
0.005	83.05	195.0	83.13	197.0	81.75	203.0
0.003	85.98	194.0	86.06	194.0	84.81	200.0
0.001	92.24	198.0	92.26	195.0	91.24	202.0
0.0007	94.34	202.0	94.31	198.0	93.37	204.0
0.0005	96.45	208.0	96.36	202.0	95.46	211.0
0.0003	99.76	224.0	99.53	211.0	98.67	223.0
0.0001	108.3	294.0	107.2	256.0	107.2	291.0

Model atmospheres (0–105 km)

June

Pressure (mb)	Geopotential height (km) 10° N	Temp. (K)	Geopotential height (km) 40° N	Temp. (K)	Geopotential height (km) 70° N	Temp. (K)
1000	0.098	299.5	0.137	290.4	0.096	277.5
950.0	0.544	296.7	0.571	289.1	0.515	276.8
900.0	1.017	294.3	1.028	287.6	0.953	275.3
850.0	1.513	291.8	1.506	286.0	1.412	273.3
700.0	3.160	283.1	3.112	278.0	2.943	265.6
500.0	5.873	267.0	5.771	261.2	5.488	250.3
400.0	7.587	256.5	7.437	249.5	7.085	238.9
300.0	9.689	241.4	9.482	234.7	9.048	227.7
200.0	12.42	219.1	12.17	219.3	11.73	226.9
100 0	16.60	198.3	16.54	213.7	16.38	229.0
70.0	18.73	203.9	18.79	215.9	18.75	229.5
50.0	20.73	209.2	20.92	218.0	20.99	230.1
30.0	23.91	218.8	24.16	222.9	24.41	231.2
10.0	31.09	238.0	31.26	237.0	31.71	243.0
7.0	33.62	246.0	33.76	244.0	34.28	248.0
5.0	36.06	252.0	36.20	251.0	36.76	255.0
3.0	39.88	260.0	40.04	263.0	40.62	266.0
1.0	48.50	272.0	48.81	276.0	49.53	283 0
0.70	51.34	271.0	51.66	273.0	52.48	280.0
0.50	54.00	268.0	54.33	269.0	55.24	276.0
0.30	57.95	259.0	58.28	256.0	59.28	265.0
0.10	65.77	227.0	65.97	225.0	67.28	234.0
0.07	68.09	216.0	68.31	216.0	69.70	224 0
0.05	70.19	208.0	70.39	208.0	71.81	213.0
0.03	73.24	201.0	73.41	198.0	74.92	198.0
0.01	79.61	196.0	79.50	180.0	80.67	166.0
0.007	81.65	195.0	81.32	176.0	82.35	158.0
0.005	83.57	195.0	83.04	172.0	83.94	150.0
0.003	86.45	192.0	85.57	167.0	86.11	147.0
0.001	92.54	187.0	90.93	171.0	90.90	156.0
0.0007	94.54	188.0	92.67	178.0	92.54	165.0
0.0005	96.44	190.0	94.49	188.0	94.15	176.0
0.0003	99.41	198.0	97.60	207.0	97.15	209.0
0.0001	106.6	243.0	105.3	263.0	105.8	327.0

September

Pressure (mb)	Geopotential height (km) 10° N	Temp. (K)	Geopotential height (km) 40° N	Temp. (K)	Geopotential height (km) 70° N	Temp. (K)
1000	0.092	299.4	0.141	291.9	0.076	274.2
950.0	0.538	296.6	0.578	289.9	0.491	272.5
900.0	1.010	294.1	1.037	287.8	0.924	270.8
850.0	1.505	291.5	1.518	285.7	1.378	269.4
700.0	3.149	282.7	3.125	278.1	2.888	262.3
500.0	5.859	266.9	5.789	261.9	5.402	247.4
400.0	7.570	256.3	7.457	250.3	6.981	236.3
300.0	9.670	241.0	9.514	235.8	8.923	224.6
200.0	12.40	218.7	12.21	219.7	11.56	222.8
100.0	16.58	199.7	16.54	211.5	16.11	224.5
70.0	18.72	205.5	18.78	214.3	18.44	224.0
50.0	20.73	211.0	20.89	217.0	20.63	223.6
30.0	23.93	217.6	24.13	221.6	23.95	222.2
10.0	31.07	235.0	31.35	236.0	31.41	232.0
7.0	33.54	241.0	33.83	240.0	33.84	237.0
5.0	35.95	247.0	36.22	246.0	36.20	242.0
3.0	39.70	258.0	39.93	256.0	39.88	252.0
1.0	48.30	272.0	48.48	269.0	48.37	270.0
0.70	51.14	272.0	51.27	266.0	51.18	271.0
0.50	53.83	269.0	53.90	260.0	53.84	268.0
0.30	57.77	259.0	57.67	248.0	57.79	260.0
0.10	65.67	233.0	65.23	221.0	65.75	235.0
0.07	68.07	225.0	67.48	214.0	68.17	225.0
0.05	70.25	218.0	69.57	208.0	70.36	216.0
0.03	73.44	209.0	72.63	201.0	73.49	204.0
0.01	80.01	200.0	78.90	190.0	79.71	183.0
0.007	82.08	199.0	80.87	188.0	81.57	179.0
0.005	84.03	199.0	82.73	186.0	83.32	175.0
0.003	87.02	199.0	85.53	187.0	85.93	173.0
0.001	93.49	207.0	91.71	198.0	91.54	176.0
0.0007	95.81	216.0	93.79	203.0	93.39	178.0
0.0005	98.00	224.0	95.87	207.0	95.19	180.0
0.0003	101.7	247.0	99.06	213.0	97.96	183.0
0.0001	110.9	338.0	106.7	250.0	104.1	188.0

Model atmospheres (0-105 km)

December

Pressure (mb)	Geopotential height (km) 10° N	Temp. (K)	Geopotential height (km) 40° N	Temp. (K)	Geopotential height (km) 70° N	Temp. (K)
1000	0.097	299.4	0.133	280.2	0.096	254.5
950.0	0.546	296.7	0.557	278.0	0.491	254.6
900.0	1.017	293.9	1.000	275.7	0.897	255.3
850.0	1.509	291.2	1.465	273.5	1.322	255.6
700.0	3.150	283.0	3.011	266.6	2.764	250.8
500.0	5.863	267.6	5.567	250.9	5.191	236.6
400.0	7.579	256.7	7.170	239.3	6.704	226.9
300.0	9.679	241.2	9.123	225.8	8.556	217.7
200.0	12.41	219.0	11.72	215.8	11.10	214.9
100.0	16.57	196.1	16.09	213.6	15.49	214.3
70.0	18.65	201.7	18.33	214.1	17.83	214.3
50.0	20.62	207.1	20.44	214.7	20.04	214.4
30.0	23.76	216.2	23.64	216.8	22.89	209.0
10.0	30.97	237.0	30.56	225.0	30.36	212.0
7.0	33.47	244.0	32.95	230.0	32.61	215.0
5.0	35.90	251.0	35.22	235.0	34.72	218.0
3.0	39.70	259.0	38.83	245.0	38.04	224.0
1.0	48.27	271.0	47.09	264.0	45.51	245.0
0.70	51.09	271.0	49.87	266.0	48.14	253.0
0.50	53.76	269.0	52.46	263.0	50.64	258.0
0.30	57.71	259.0	56.34	256.0	54.52	256.0
0.10	65.49	224.0	64.28	235.0	62.47	238.0
0.07	67.77	214.0	66.68	229.0	64.97	232.0
0.05	69.87	207.0	68.93	223.0	67.20	228.0
0.03	72.90	202.0	72.18	216.0	70.55	222.0
0.01	79.36	201.0	78.96	204.0	77.57	215.0
0.007	81.46	200.0	81.06	203.0	79.82	214.0
0.005	83.44	200.0	83.04	202.0	81.92	214.0
0.003	86.37	193.0	86.07	202.0	85.12	215.0
0.001	92.43	185.0	92.64	206.0	92.11	218.0
0.0007	94.40	186.0	94.83	208.0	94.42	219.0
0.0005	96.30	189.0	96.95	210.0	96.60	218.0
0.0003	99.27	199.0	100.2	216.0	99.93	216.0
0.0001	106.7	257.0	107.9	248.0	107.1	215.0

Appendices

March

Geopotential height (km)	Pressure (mb) 10° N	Temp. (K)	Pressure (mb) 40° N	Temp. (K)	Pressure (mb) 70° N	Temp (K)
0	1011.0	299.6	1018.0	281.2	1014.0	253.0
1	901.7	294.1	899.1	275.7	889.0	254.6
2	802.2	289.0	793.5	271.0	776.7	252.5
3	712.7	283.8	699.6	266.3	677.8	248.6
4	630.1	278.2	613.1	260.1	589.2	242.8
5	556.6	272.4	537.4	253.9	512.1	237.0
6	491.2	266.5	469.3	247.1	442.3	230.9
7	431.3	260.1	408.2	239.8	380.8	225.3
8	377.6	253.4	352.8	232.8	326.5	221.2
9	329.2	246.2	304.5	225.9	279.4	218.8
10	286.0	238.8	260.8	222.1	238.7	218.5
11	246.5	230.7	223.2	218.7	203.9	218.3
12	212.5	222.6	190.8	216.3	174.6	218.5
13	181.2	216.0	163.0	216.1	149.5	218.8
14	153.3	210.4	139.2	215.8	128.1	219.0
15	129.8	204.8	118.9	215.6	109.7	219.3
16	109.8	199.3	101.5	215.4	94.09	219.5
17	92.80	197.2	86.65	215.5	80.89	219.7
18	78.24	199.5	73.95	215.7	69.54	219.9
19	65.96	201.7	63.11	215.8	59.78	220.1
20	55.61	204.0	53.86	216.0	51.39	220.3
21	47.02	206.8	45.98	216.5	43.55	219.1
22	39.94	210.5	39.26	217.4	36.77	217.5
23	33.93	214.2	33.53	218.3	31.06	215.9
24	28.90	217.5	28.67	219.2	26.50	216.2
25	24.81	219.5	24.59	220.4	22.67	216.9
30	11.4	233.0	11.0	224.0	10.4	226.0
35	5.56	248.0	5.19	237.0	4.94	231.0
40	2.85	262.0	2.59	252.0	2.37	237.0
45	1.50	272.0	1.34	264.0	1.18	247.0
50	0.808	273.0	0.705	267.0	0.595	255.0
55	0.427	264.0	0.368	260.0	0.305	252.0
60	0.218	242.0	0.188	245.0	0.153	240.0
65	0.104	222.0	0.917×10^{-1}	233.0	0.733×10^{-1}	227.0
70	0.470×10^{-1}	208.0	0.433×10^{-1}	221.0	0.339×10^{-1}	217.0
75	0.202×10^{-1}	200.0	0.196×10^{-1}	211.0	0.153×10^{-1}	210.0
80	0.858×10^{-2}	196.0	0.860×10^{-2}	202.0	0.668×10^{-2}	206.0
85	0.355×10^{-2}	194.0	0.361×10^{-2}	195.0	0.290×10^{-2}	200.0
90	0.149×10^{-2}	195.0	0.150×10^{-2}	192.0	0.123×10^{-2}	200.0
95	0.630×10^{-3}	204.0	0.625×10^{-3}	199.0	0.538×10^{-3}	209.0
100	0.289×10^{-3}	225.0	0.279×10^{-3}	213.0	0.250×10^{-3}	232.0
105	0.146×10^{-3}	263.0	0.134×10^{-3}	239.0	0.129×10^{-3}	270.0

172

Model atmospheres (0–105 km)

June

Geopotential height (km)	Pressure (mb) 10° N	Temp. (K)	Pressure (mb) 40° N	Temp. (K)	Pressure (mb) 70° N	Temp. (K)
0	1011.0	300.1	1016.0	290.8	1012.0	277.6
1	901.8	294.3	903.0	287.7	894.7	275.1
2	802.6	289.2	800.7	283.5	788.9	270.3
3	713.3	283.9	709.5	278.5	694.7	265.2
4	630.7	278.1	625.6	272.3	608.7	259.2
5	557.2	272.1	551.2	266.0	533.3	253.2
6	491.8	266.2	484.9	259.6	465.5	246.6
7	431.8	260.1	424.1	252.5	404.8	239.5
8	378.0	253.5	369.5	245.4	349.8	233.6
9	329.7	246.3	321.0	238.1	302.1	227.9
10	286.5	238.8	277.4	231.7	259.8	227.4
11	247.0	230.7	238.5	225.9	223.4	227.1
12	212.9	222.5	205.0	220.2	192.1	227.0
13	181.8	216.2	175.2	218.2	165.5	227.4
14	154.0	211.2	149.6	216.9	142.6	227.9
15	130.4	206.2	127.6	215.6	122.8	228.3
16	110.5	201.3	108.9	214.4	105.8	228.8
17	93.55	199.3	92.98	214.1	91.05	229.1
18	79.10	201.9	79.36	215.1	78.34	229.3
19	66.88	204.6	67.73	216.1	67.40	229.6
20	56.55	207.2	57.80	217.1	58.00	229.8
21	47.90	210.0	49.34	218.1	49.90	230.1
22	40.80	213.0	42.15	219.6	42.99	230.4
23	34.74	216.0	36.01	221.1	37.03	230.7
24	29.61	218.9	30.77	222.6	31.90	231.0
25	25.47	220.9	26.47	224.3	27.54	231.9
30	11.7	235.0	12.0	233.0	12.7	239.0
35	5.77	250.0	5.87	248.0	6.34	249.0
40	2.95	261.0	3.01	263.0	3.24	264.0
45	1.55	269.0	1.60	275.0	1.74	278.0
50	0.829	273.0	0.862	275.0	0.944	283.0
55	0.440	267.0	0.459	267.0	0.515	277.0
60	0.229	253.0	0.238	250.0	0.273	262.0
65	0.113	230.0	0.116	229.0	0.140	243.0
70	0.516×10^{-1}	209.0	0.534×10^{-1}	209.0	0.667×10^{-1}	222.0
75	0.222×10^{-1}	198.0	0.227×10^{-1}	193.0	0.296×10^{-1}	197.0
80	0.933×10^{-2}	196.0	0.907×10^{-2}	179.0	0.115×10^{-1}	169.0
85	0.388×10^{-2}	193.0	0.337×10^{-2}	168.0	0.391×10^{-2}	148.0
90	0.159×10^{-2}	188.0	0.121×10^{-2}	167.0	0.122×10^{-2}	151.0
95	0.645×10^{-3}	188.0	0.460×10^{-3}	191.0	0.433×10^{-3}	186.0
100	0.272×10^{-3}	201.0	0.207×10^{-3}	224.0	0.195×10^{-3}	244.0
105	0.126×10^{-3}	229.0	0.104×10^{-3}	260.0	0.108×10^{-3}	316.0

September

Geopotential height (km)	Pressure (mb) 10° N	Temp. (K)	Pressure (mb) 40° N	Temp. (K)	Pressure (mb) 70° N	Temp (K)
0	1011.0	299.9	1017.0	292.5	1009.0	274.5
1	901.0	294.1	903.9	287.9	891.4	270.5
2	801.7	288.8	801.9	283.4	784.7	266.4
3	712.4	283.5	710.7	278.7	689.6	261.6
4	629.8	277.7	626.8	272.7	603.2	255.7
5	556.3	271.9	552.4	266.7	527.6	249.7
6	490.9	266.0	486.1	260.4	459.5	243.2
7	430.9	259.8	425.2	253.4	398.9	236.1
8	377.1	253.1	370.7	246.4	344.0	230 1
9	328.8	245.8	322.4	239.4	296.5	224 5
10	285.7	238.3	278.8	232.9	254.2	223.8
11	246.2	230.1	239.9	226.9	217.9	223.1
12	212.3	221.9	206.4	220.9	187 0	222 9
13	181.1	215.9	176.2	218.2	160.6	223.3
14	153.4	211.4	150.2	216 3	137.9	223.7
15	130.0	206.9	128.0	214.4	118.5	224.0
16	110.2	202.3	109.1	212.5	101.7	224.4
17	93.28	200.8	92.98	212.0	87.27	224 3
18	78.93	203 5	79.27	213.3	74.85	224.1
19	66.78	206.2	67.58	214.6	64.20	223.9
20	56.51	209.0	57.61	215.8	55.07	223.7
21	47.90	211.5	49.13	217.1	47.23	223.4
22	40.83	213.6	41.96	218.5	40.49	223.0
23	34.80	215.6	35.85	220.0	34.72	222.6
24	29.68	217.7	30.62	221.4	29.77	222.2
25	25.51	219.7	26.30	222.7	25.56	222.8
30	11.7	232.0	12.2	233.0	12.3	229.0
35	5.69	245.0	5.92	243.0	5.91	239.0
40	2.89	259.0	2.97	257.0	2.95	252.0
45	1.51	269.0	1.56	267.0	1.53	266.0
50	0.808	273.0	0.824	269.0	0.813	272.0
55	0.431	267.0	0.433	257.0	0.431	266.0
60	0.223	252.0	0.217	239.0	0.224	255.0
65	0.110	235.0	0.104	222.0	0.112	238.0
70	0.520×10^{-1}	219.0	0.466×10^{-1}	207.0	0.530×10^{-1}	218.0
75	0.232×10^{-1}	206.0	0.200×10^{-1}	197.0	0.232×10^{-1}	199.0
80	0.100×10^{-1}	200.0	0.820×10^{-2}	189.0	0.946×10^{-2}	182.0
85	0.424×10^{-2}	199.0	0.331×10^{-2}	186.0	0.360×10^{-2}	173.0
90	0.181×10^{-2}	201.0	0.134×10^{-2}	194.0	0.135×10^{-2}	174.0
95	0.793×10^{-3}	213.0	0.575×10^{-3}	205.0	0.518×10^{-3}	180.0
100	0.378×10^{-3}	236.0	0.260×10^{-3}	216.0	0.208×10^{-3}	185.0
105	0.198×10^{-3}	277.0	0.126×10^{-3}	238.0	0.862×10^{-4}	190.0

Model atmospheres (0–105 km)

December

Geopotential height (km)	Pressure (mb) 10° N	Temp. (K)	Pressure (mb) 40° N	Temp. (K)	Pressure (mb) 70° N	Temp. (K)
0	1011.0	299.9	1016.0	280.9	1013.0	254.4
1	901.8	294.0	900.0	275.7	887.6	255.3
2	802.0	288.7	794.8	271.1	775.8	253.3
3	712.5	283.7	701.0	266.6	677.5	249.4
4	630.0	278.1	614.5	260.5	589.8	243.5
5	556.5	272.5	538.7	254.3	513.4	237.7
6	491.2	266.7	470.8	247.7	443.8	231.4
7	431.3	260.3	409.6	240.5	382.0	225.4
8	377.6	253.6	354.0	233.5	327.1	220.4
9	329.2	246.2	305.5	226.6	279.5	217.2
10	286.0	238.5	261.7	222.4	238.3	216.1
11	246.5	230.4	223.9	218.5	203.2	215.0
12	212.5	222.3	191.4	215.6	173.5	214.7
13	181.2	215.7	163.3	215.1	148.2	214.6
14	153.4	210.2	139.3	214.6	126.6	214.5
15	129.9	204.7	118.9	214.1	108.1	214.3
16	110.0	199.2	101.4	213.6	92.57	214.3
17	92.92	197.2	86.48	213.8	79.49	214.3
18	78.29	199.9	73.75	214.0	68.26	214.3
19	65.97	202.7	62.89	214.3	58.61	214.3
20	55.58	205.4	53.63	214.5	50.33	214.4
21	46.99	208.2	45.72	215.0	42.12	212.5
22	39.94	211.1	38.97	215.7	35.21	210.7
23	33.95	214.0	33.22	216.3	29.49	209.0
24	28.92	216.7	28.37	217.1	25.05	209.1
25	24.83	218.7	24.36	218.2	21.28	209.3
30	11.5	235.0	10.9	224.0	10.6	211.0
35	5.64	249.0	5.16	235.0	4.79	218.0
40	2.89	260.0	2.55	249.0	2.22	227.0
45	1.51	267.0	1.31	263.0	1.07	244.0
50	0.804	272.0	0.688	266.0	0.544	258.0
55	0.427	266.0	0.359	260.0	0.281	255.0
60	0.221	252.0	0.183	247.0	0.142	244.0
65	0.108	226.0	0.899×10^{-1}	233.0	0.697×10^{-1}	232.0
70	0.489×10^{-1}	207.0	0.424×10^{-1}	220.0	0.326×10^{-1}	223.0
75	0.210×10^{-1}	200.0	0.191×10^{-1}	211.0	0.150×10^{-1}	217.0
80	0.898×10^{-2}	201.0	0.838×10^{-2}	203.0	0.680×10^{-2}	214.0
85	0.381×10^{-2}	197.0	0.359×10^{-2}	201.0	0.306×10^{-2}	215.0
90	0.157×10^{-2}	187.0	0.155×10^{-2}	204.0	0.138×10^{-2}	217.0
95	0.630×10^{-3}	187.0	0.682×10^{-3}	208.0	0.640×10^{-3}	219.0
100	0.266×10^{-3}	203.0	0.311×10^{-3}	215.0	0.297×10^{-3}	215.0
105	0.125×10^{-3}	240.0	0.149×10^{-3}	232.0	0.138×10^{-3}	212.0

6 MEAN REFERENCE ATMOSPHERE 110–500 km

Geometric height (km)	Temp. (K)	Log pressure (mb)	Mean mol. wt.
110	244	− 4.121	26.56
120	335	− 4.574	25.45
130	445	− 4.892	24.48
140	549	− 5.130	23.64
150	635	− 5.322	22.91
160	703	− 5.487	22.25
170	756	− 5.633	21.65
180	798	− 5.767	21.10
190	832	− 5.890	20.59
200	859	− 6.006	20.13
250	940	− 6.512	18.33
300	973	− 6.947	17.21
350	987	− 7.345	16.52
400	993	− 7.721	16.02
450	996	− 8.070	15.54
500	997	− 8.417	14.94

From COSPAR (1972): see chapter 5 for more information.

7 THE PLANCK FUNCTION

$$B_{\tilde{\nu}} = \frac{c_1 \tilde{\nu}^3}{\exp(c_2 \tilde{\nu}/T) - 1}$$

where $B_{\tilde{\nu}}$ is radiance (in $W\,m^{-2}\,sr^{-1}(cm^{-1})^{-1}$) of black body at T K and $\tilde{\nu}$ wavenumbers (cm^{-1}). c_1 and c_2 are known as first and second radiation constants and have the values

$$c_1 = 1.1911 \times 10^{-8}\,W\,m^{-2}\,sr^{-1}\,(cm^{-1})^{-4}$$

$$c_2 = 1.439\,K\,(cm^{-1})^{-1}$$

$$B_\lambda = \frac{c_1}{\lambda^5(\exp(c_2/\lambda T) - 1)}$$

where B_λ is radiance in $W\,m^{-2}\,sr^{-1}\,cm^{-1}$ of black body at T K and wavelength λ cm. c_1 and c_2 have the same values as above.

Another useful quantity is

$$D = \frac{\int_{\tilde{\nu}}^{\infty} B_{\tilde{\nu}}\,d\tilde{\nu}}{\int_0^{\infty} B_{\tilde{\nu}}\,d\tilde{\nu}}$$

A series evaluation of D suitable for numerical computation is

$$D = \frac{15}{\pi^4} \sum_{m=1}^{\infty} m^{-4} \exp(-mv)[\{(mv + 3)mv + 6\}mv + 6]$$

176

Solar radiation

For small values of v, $v < 2\pi$, the following series converges more rapidly

$$D = 1 - \frac{15}{\pi^4} v^3 \left(\frac{1}{3} - \frac{v}{8} + \frac{v^2}{60} - \frac{v^4}{5040} + \frac{v^6}{272\,160} - \frac{v^8}{13\,305\,600} \right)$$

where

$$v = \frac{c_2 \tilde{v}}{T}$$

The wavenumber $\tilde{v}_m$ cm^{-1} of maximum $B_{\tilde{v}}$ is given by Wien's displacement law which is

$$\tilde{v}_m = 1.9609T$$

The wavelength λ_m cm of maximum B_λ is given by

$$\lambda_m = 0.289\,79/T.$$

Reference Pivovonsky & Nagel (1961).

8 SOLAR RADIATION

Solar constant (i.e. mean value of total solar radiation incident on surface just outside earth's atmosphere normal to solar beam) $= 1370\ \mathrm{W\,m^{-2}}$.

Table A8. *Solar Spectral Irradiance* (after Thekaekara, 1973)

λ = wavelength, μm
I_λ = solar spectral irradiance averaged over small bandwidth centred at λ (W m^{-2} nm^{-1})
$D_{0-\lambda}$ = percentage of the solar constant associated with wavelengths shorter than λ

λ	I_λ	$D_{0-\lambda}$	λ	I_λ	$D_{0-\lambda}$	λ	I_λ	$D_{0-\lambda}$
0.115	7×10^{-6}	1×10^{-4}	0.43	1.660	12.47	0.90	0.902	63.37
0.14	3×10^{-5}	5×10^{-4}	0.44	1.833	13.73	1.00	0.757	69.49
0.16	2.3×10^{-4}	6×10^{-4}	0.45	2.031	15.14	1.2	0.491	78.40
0.18	0.00127	1.6×10^{-3}	0.46	2.092	16.65	1.4	0.341	84.33
0.20	0.0108	8.1×10^{-3}	0.47	2.059	18.17	1.6	0.248	88.61
0.22	0.0582	0.05	0.48	2.100	19.68	1.8	0.161	91.59
0.23	0.0675	0.10	0.49	1.975	21.15	2.0	0.104	93.49
0.24	0.0638	0.14	0.50	1.966	22.60	2.2	0.080	94.83
0.25	0.0718	0.19	0.51	1.906	24.01	2.4	0.063	95.86
0.26	0.132	0.27	0.52	1.856	25.38	2.6	0.049	96.67
0.27	0.235	0.41	0.53	1.865	26.74	2.8	0.039	97.31
0.28	0.225	0.56	0.54	1.805	28.08	3.0	0.031	97.83
0.29	0.488	0.81	0.55	1.747	29.38	3.2	0.0229	98.22
0.30	0.520	1.21	0.56	1.716	30.65	3.4	0.0168	98.50
0.31	0.698	1.66	0.57	1.734	31.91	3.6	0.0137	98.72
0.32	0.840	2.22	0.58	1.737	33.18	3.8	0.0112	98.91
0.33	1.072	2.93	0.59	1.721	34.44	4.0	0.0096	99.06
0.34	1.087	3.72	0.60	1.687	35.68	4.5	0.0060	99.34
0.35	1.107	4.52	0.62	1.622	38.10	5.0	0.0038	99.51
0.36	1.081	5.32	0.64	1.563	40.42	6.0	0.0018	99.72
0.37	1.190	6.15	0.66	1.505	42.66	7.0	0.0010	99.82
0.38	1.134	7.00	0.68	1.445	44.81	8.0	6.0×10^{-4}	99.88
0.39	1.112	7.82	0.70	1.386	46.88	10.0	2.5×10^{-4}	99.94
0.40	1.447	8.73	0.72	1.331	48.86	15.0	4.9×10^{-5}	99.98
0.41	1.773	9.92	0.75	1.251	51.69	20.0	1.5×10^{-5}	99.99
0.42	1.770	11.22	0.80	1.123	56.02	50.0	4×10^{-7}	100.00

177

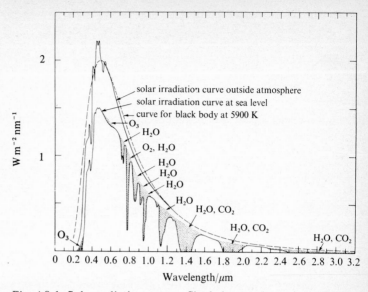

Fig. A8.1. Solar radiation curves. Shaded areas show absorption by vertical path of whole atmosphere by constituents shown (from Air Force Cambridge Research Laboratories, 1965).

9 ABSORPTION OF SOLAR RADIATION BY OXYGEN AND OZONE

Table A9.1. *Absorption cross-sections per molecule of* O_2 *and* O_3 *in* cm^2

λ nm	O_2	O_3
150	16.8×10^{-18}	4.7×10^{-19}
160	5.32×10^{-18}	1.1×10^{-18}
170	1.86×10^{-18}	8.2×10^{-19}
180	15.0×10^{-24}	7.4×10^{-19}
190	15.4×10^{-24}	5.1×10^{-19}
200	13.0×10^{-24}	2.9×10^{-19}
210	9.6×10^{-24}	4.5×10^{-19}
220	6.4×10^{-24}	1.9×10^{-18}
230	3.1×10^{-24}	4.6×10^{-18}
240	1.0×10^{-24}	8.0×10^{-18}
250	3.0×10^{-25}	10.8×10^{-18}
260		10.7×10^{-18}
270		7.8×10^{-18}
280		3.7×10^{-18}
290		1.3×10^{-18}
300		4×10^{-19}
310		1×10^{-19}
320		3×10^{-20}
330		7×10^{-21}
340		2×10^{-21}
450		0.2×10^{-21}
500		1.0×10^{-21}
550		3.3×10^{-21}
600		4.7×10^{-21}
650		2.7×10^{-21}
700		0.9×10^{-21}
750		0.3×10^{-21}

Values for O_2 are from Ditchburn & Young (1962), for O_3 below 300 nm from Inn & Tanaka (1953), above 300 nm from Vigroux (1953). The cross-sections for O_2 include the effect of Rayleigh scattering.

Appendices

Table A9.2. *Solar absorption due to ozone by all ultraviolet and visible bands for various path lengths along solar beam.*

Ozone path length (cm STP)	Rate of absorption of solar radiation (mW cm^{-2})
0.0001	0.0195
0.0005	0.0928
0.0010	0.1772
0.0050	0.6530
0.0100	0.9766
0.0500	1.848
0.1000	2.345
0.2000	3.005
0.3000	3.523
0.4000	3.979
0.5000	4.400
0.6000	4.796
0.7000	5.175
0.8000	5.541
0.9000	5.895
1.0000	6.241
2.0000	9.360
3.0000	12.096
4.0000	14.564
5.0000	16.818
6.0000	18.887
7.0000	20.795
8.0000	22.559
9.0000	24.195
10.0000	25.714

From Kennedy (1964).

Absorption by near infrared O_2 bands (after Houghton, 1963)

For atmospheric paths where collision broadening is dominant, absorption by the O_2 bands at 0.76 μm and 1.27 μm follows the square-root law (4.37) with $\Sigma (s_i\gamma_{0i})^{1/2} = 2.3$ cm^{-1}(g cm^{-2})$^{-1/2}$ for 0.76 μm band and $= 0.07$ cm^{-1}(g cm^{-2})$^{-1/2}$ for 1.27 μm band.

10 SPECTRAL BAND INFORMATION

The following tables list values of Σs_i (columns S) and $\Sigma (s_i\gamma_{0i})^{1/2}$ (columns R) at different temperatures over different wavenumber intervals of various infrared bands of water vapour, carbon dioxide and ozone, where s_i is the strength of the ith line in cm^{-1}(g cm^{-2})$^{-1}$ and γ_{0i} is the collision broadened half-width in cm^{-1} of the ith line at standard pressure (1013 mb). The data are from McClatchey *et al.* (1973). More explanation is given in chapter 4.

In the tables, each number is given by a number between 1 and 10 expressed to four significant figures followed by an exponent of ten e.g. the value for S at 220 K for water vapour between 0 and 25 cm^{-1} is 3.615 × 10^3.

180

Spectral band information

Water vapour

Wavenumber interval	220 K S		R		260 K S		R		300 K S		R	
0–25	3.615	3	3.191	1	2.487	3	2.677	1	1.801	3	2.309	1
25–50	3.706	4	2.097	2	2.971	4	1.829	2	2.426	4	1.619	2
50–75	1.187	5	4.251	2	9.374	4	3.797	2	7.622	4	3.433	2
75–100	2.176	5	5.404	2	1.768	5	4.944	2	1.469	5	4.563	2
100–125	1.973	5	4.853	2	1.854	5	4.772	2	1.711	5	4.642	2
125–150	2.290	5	4.858	2	2.063	5	4.698	2	1.863	5	4.529	2
150–175	2.955	4	4.867	2	2.754	5	4.847	2	2.552	5	4.783	2
175–200	7.944	5	2.498	2	9.435	4	2.804	2	1.078	5	3.044	2
200–225	2.450	5	4.231	2	2.460	5	4.259	2	2.395	5	4.230	2
225–250	1.259	4	2.649	2	1.407	5	2.844	2	1.494	5	2.981	2
250–275	7.900	4	1.600	2	9.516	4	1.760	2	1.047	5	1.867	2
275–300	6.289	4	2.085	2	8.495	4	2.368	2	1.023	5	2.563	2
300–325	4.341	4	1.581	2	6.570	4	1.883	2	8.630	4	2.119	2
325–350	1.870	3	1.085	2	3.193	4	1.346	2	4.645	4	1.562	2
350–375	9.650	3	8.448	1	1.822	4	1.126	2	2.934	4	1.383	2
375–400	4.413	3	6.273	1	9.153	3	8.692	1	1.638	4	1.110	1
400–425	3.004	2	3.724	1	5.209	3	5.274	1	8.319	3	6.876	1
425–450	7.356	3	2.322	1	1.337	3	3.232	1	2.700	3	4.365	1
450–475	1.475	2	3.374	1	2.598	3	4.652	1	4.084	3	5.994	1
475–500	3.264	2	9.622	0	4.893	3	1.413	0	8.023	3	2.023	1
500–525	5.422	2	1.654	1	1.079	3	2.435	1	1.856	2	3.340	1
525–550	3.074	2	1.088	1	5.415	2	1.453	1	8.219	2	1.870	1
550–575	1.358	2	7.994	0	2.318	2	1.083	1	3.601	2	1.429	1
575–600	1.635	2	8.508	0	3.686	2	1.272	0	6.631	2	1.713	1
600–625	4.782	1	3.981	0	1.026	2	5.926	0	1.955	2	8.264	0
625–650	5.678	1	4.855	0	1.379	2	7.300	0	2.681	2	9.932	0
650–675	2.575	1	3.175	0	4.857	1	4.496	0	8.300	1	5.975	0
675–700	1.779	1	2.660	0	5.154	1	4.345	0	1.124	1	6.186	0
700–725	9.761	0	1.876	0	2.430	1	3.040	0	5.154	1	4.427	0
725–750	5.737	0	1.524	0	1.543	1	2.453	0	3.479	1	3.555	0
750–775	1.764	0	7.205	−1	4.776	0	1.196	0	1.145	1	1.817	0
775–800	5.477	0	1.567	0	1.410	1	2.536	0	2.939	1	3.621	0

Water vapour (continued)

Temperature Wavenumber interval	220 K S		220 K R		260 K S		260 K R		300 K S		300 K R	
1200 1250	1.682	1	5.706	0	3.394	1	8.364	0	6.166	1	1.166	1
1250 1300	8.281	1	1.562	1	1.760	2	2.274	1	3.411	2	3.133	1
1300 1350	6.019	2	4.324	1	1.271	3	5.869	1	2.253	3	7.493	1
1350 1400	5.161	3	9.781	1	8.007	3	1.206	2	1.098	4	1.408	1
1400 1450	1.043	4	1.313	2	1.357	4	1.499	2	1.629	4	1.653	2
1450 1500	3.198	4	2.609	2	3.526	4	2.763	2	3.739	4	2.888	2
1500 1550	6.466	4	4.780	2	6.573	4	5.012	2	6.640	4	5.210	2
1550 1600	4.774	4	2.673	2	4.392	4	2.559	2	4.062	4	2.462	2
1600 1650	4.252	4	2.640	2	3.667	4	2.450	2	3.209	4	2.306	2
1650 1700	8.557	4	4.408	2	7.739	4	4.220	2	7.057	4	4.059	2
1700 1750	4.263	4	3.173	2	4.412	4	3.252	2	4.463	4	3.298	2
1750 1800	2.301	4	1.906	2	2.553	4	2.062	2	2.781	4	2.192	2
1800 1850	8.472	3	1.021	2	9.912	3	1.150	2	1.134	4	1.270	2
1850 1900	2.955	3	5.064	1	3.575	3	5.746	1	4.097	3	6.408	1
1900 1950	1.896	3	4.285	1	2.581	3	5.030	1	3.186	3	5.690	1
1950 2000	4.045	2	2.033	1	6.707	2	2.546	1	9.549	2	3.028	1
2000 2050	1.109	2	1.084	1	2.157	2	1.511	1	3.504	2	1.948	1
2050 2100	3.618	1	6.195	0	7.955	1	9.316	0	1.425	2	1.274	1
2800 2900	5.145	0	9.590	0	6.051	0	1.023	1	7.034	0	1.084	1
2900 3000	2.932	1	1.207	1	4.680	1	1.543	1	6.665	1	1.874	1
3000 3100	4.137	2	4.243	1	4.351	2	4.425	1	4.472	2	4.574	1
3100 3200	5.370	2	4.656	1	4.888	2	4.682	1	4.537	2	4.785	1
3200 3300	7.944	2	6.302	1	7.602	2	6.460	1	7.371	2	6.683	1
3300 3400	3.752	2	4.683	1	4.641	2	5.672	1	5.745	2	6.672	1
3400 3500	7.461	2	7.408	1	1.105	3	9.470	1	1.559	3	1.164	2
3500 3600	8.166	3	2.762	2	1.198	4	3.267	2	1.623	4	3.710	2
3600 3700	6.589	4	5.927	2	6.819	4	5.947	2	6.873	4	5.926	2
3700 3800	8.352	4	7.355	2	7.880	4	7.280	2	7.525	4	7.222	2
3800 3900	1.084	5	7.192	2	1.025	5	7.026	2	9.669	4	6.854	2
3900 4000	7.588	3	1.718	2	1.112	4	2.109	2	1.483	4	2.461	2
4000 4100	3.194	2	2.943	1	3.707	2	3.378	1	4.262	2	3.877	1
4100 4200	4.760	1	1.211		3.707							

Freq	n	v1	n	v2	n	v3	n	v4	n	v5	n	v6
5000	0	3.722	0	5.244	0	6.551	0	6.874	0	1.126	0	8.433
5100	1	4.905	1	1.678	3	7.132	1	2.052	1	1.029	1	2.455
5200	2	8.698	1	7.783	3	1.270	2	9.159	2	1.692	2	1.040
5300	3	7.303	2	1.779	3	7.231	2	1.755	2	7.053	2	1.722
5400	4	1.170	2	2.295	4	1.069	2	2.254	2	9.958	2	2.228
5500	4	1.008	2	2.080	4	1.031	2	2.082	2	1.031	2	2.065
5600	2	5.709	1	5.023	2	9.270	1	6.271	1	1.343	1	7.381
5700	1	5.333	1	1.261	1	6.648	1	1.552	1	8.475	1	1.885
5800	1	1.016	0	4.586	1	1.410	0	5.517	0	1.790	0	6.341
6700	1	2.291	1	1.123	1	2.970	1	1.308	1	3.802	1	1.485
6800	2	3.071	1	4.013	1	3.454	1	4.328	2	3.748	2	4.573
6900	2	5.157	1	5.203	2	4.634	1	5.106	2	4.246	2	5.082
7000	2	7.808	1	7.108	2	7.838	1	7.610	2	8.108	2	8.122
7100	3	1.014	1	9.688	3	1.453	2	1.127	3	1.887	3	1.251
7200	3	5.269	2	1.660	3	5.367	2	1.684	3	5.362	3	1.695
7300	3	8.422	2	2.167	3	7.630	2	2.078	3	7.016	3	2.000
7400	3	8.410	2	2.191	3	8.462	2	2.209	3	8.403	3	2.202
7500	2	2.821	1	4.518	2	4.046	1	5.205	2	5.630	2	5.883
7600	1	8.358	1	2.354	1	8.864	1	2.484	1	9.295	1	2.585
7700	1	2.571	0	9.890	1	3.108	1	1.113	1	3.581	1	1.209
8200	0	4.875	0	5.427	0	5.328	0	5.704	0	5.681	0	5.903
8300	1	1.001	0	7.840	0	9.111	0	7.571	0	8.405	0	7.370
8400	1	1.586	0	9.678	1	1.487	0	9.590	0	1.410	0	9.523
8500	0	7.987	0	8.424	1	1.183	1	1.021	1	1.722	1	1.201
8600	2	1.326	1	2.629	2	1.700	1	2.989	2	2.020	2	3.266
8700	2	3.629	1	3.702	2	3.358	1	3.591	2	3.104	2	3.500
8800	2	8.407	1	6.932	2	7.732	1	6.740	2	7.189	2	6.584
8900	2	3.489	1	4.387	2	3.951	1	4.639	2	4.300	2	4.795
9000	1	2.886	1	1.530	1	3.189	1	1.622	1	3.703	1	1.717
9100	0	9.974	0	7.007	1	1.050	1	7.366	1	1.095	1	7.587
9200	0	3.973	0	3.162	0	4.662	0	3.462	0	5.182	0	3.663
10300	1	1.679	0	8.723	1	1.685	1	8.976	0	1.686	0	9.221
10400	1	3.038	1	1.411	1	3.123	1	1.487	1	3.381	1	1.559
10500	1	8.384	1	2.540	1	9.901	1	2.694	1	1.105	1	2.782
10600	2	2.084	1	3.420	2	2.061	1	3.435	1	2.028	1	3.424
10700	2	3.267	1	4.055	2	2.889	1	3.812	1	2.578	1	3.602

Water vapour (continued)

Temperature / Wavenumber interval	220 K S	e	220 K R	e	260 K S	e	260 K R	e	300 K S	e	300 K R	e
10700	1.252	2	2.719	1	1.466	2	2.993	1	1.646	2	3.204	1
10800	1.165	1	9.701	0	1.286	1	1.004	1	1.389	1	1.021	1
10900	2.751	1	1.556	1	2.736	1	1.559	1	2.707	1	1.549	1
11000	3.378	1	1.276	1	3.031	1	1.213	1	2.753	1	1.156	1
11100	2.448	1	1.008	1	2.606	1	1.057	1	2.717	1	1.087	1

Carbon dioxide

Wavenumber interval	220 K S	e	220 K R	e	260 K S	e	260 K R	e	300 K S	e	300 K R	e
425	6.891	−5	6.845	−3	7.322	−4	2.206	−2	4.060	−3	5.116	−2
450	1.681	−3	7.073	−2	1.293	−2	1.921	−1	5.681	−2	3.945	−1
475	9.310	−4	7.316	−2	7.474	−3	2.069	−1	3.416	−2	4.395	−1
500	1.952	−2	2.870	−1	1.029	−1	6.952	−1	3.502	−1	1.337	0
525	2.785	−1	1.215	0	1.086	0	2.514	0	2.945	0	4.375	0
550	5.495	−1	2.404	0	2.684	0	5.438	0	8.844	0	1.009	1
575	5.331	1	1.958	1	1.324	2	3.365	1	2.609	2	5.118	1
600	5.196	2	5.804	1	9.797	2	8.290	1	1.562	3	1.099	2
625	7.778	3	2.084	2	9.657	3	2.530	2	1.149	4	3.015	2
650	8.746	4	7.594	2	8.597	4	8.171	2	8.526	4	8.908	2
675	2.600	4	2.635	2	2.693	4	3.049	2	2.776	4	3.547	2
700	1.232	3	8.387	1	2.339	3	1.237	2	3.756	3	1.684	2
725	2.042	3	2.852	1	4.099	2	4.572	1	6.906	2	6.670	1
750	7.278	0	6.239	0	2.693	1	1.328	1	7.086	1	2.349	1
775	1.337	0	2.765	0	4.864	0	5.651	0	1.283	1	9.858	0
800	3.974	−1	8.897	−1	1.482	0	1.805	0	3.866	0	3.065	0
825	1.280	−2	3.198	−1	9.151	−2	8.611	−1	3.878	−1	1.779	−1
850	2.501	−3	1.506	−1	1.948	−2	4.394	−1	9.093	−2	9.715	−1
875	3.937	−3	1.446	−1	2.629	−2	4.097	−1	1.211	−1	9.027	−1
900	2.320	−2	3.543	−1	1.539	−1	9.242	−1	6.447	−1	1.900	−1
925	2.859	−1	8.677	−1	1.249	−1	1.912	0	3.691	0	3.425	0
950	4.707	−1	1.024	0	1.699	0	1.948	0	4.290	0	3.150	0
975	1.906	−1	6.602	−1	8.960	−1	1.431	0	2.791	0	2.519	0
1000	4.418	−2	2.610	−1	2.257	−1	5.667	−1	7.527	−1	1.024	0
1025	3.458	−1	7.736	−1	1.338	0	1.629	0	3.600	0	2.862	0

Spectral band information

ν (low)	ν (high)	val 1	exp 1	val 2	exp 2	val 3	exp 3	val 4	exp 4	val 5	exp 5	val 6	exp 6
1050	1075	1.086	0	2.036	0	3.865	0	3.971	0	9.836	0	6.537	0
1075	1100	2.904	−2	5.006	−1	1.792	−1	1.196	0	7.142	−1	2.307	0
1800	1825	4.192	−5	5.444	−3	2.041	−4	1.155	−2	6.397	−4	1.976	−2
1825	1850	3.066	−3	7.370	−2	1.104	−2	1.329	−1	2.887	−2	2.064	−1
1850	1875	4.567	−2	4.378	−1	9.630	−2	6.529	0	1.739	−1	9.040	−1
1875	1900	3.064	−1	1.604	0	6.218	−1	2.186	0	1.102	0	2.829	0
1900	1925	5.309	0	4.130	0	5.982	0	4.660	0	6.603	0	5.160	0
1925	1950	3.358	0	3.153	0	3.301	0	3.220	0	3.273	−1	3.328	0
1950	1975	3.763	−1	7.438	−1	5.232	−1	9.260	−1	6.793	−2	1.112	−1
1975	2000	1.557	−1	2.048	−1	4.208	−2	3.207	−1	8.888	−1	4.495	−1
2000	2025	1.585	−1	5.893	−1	1.928	−1	6.977	0	2.422	−1	8.162	−1
2025	2050	3.456	0	2.788	0	5.657	0	3.290	0	8.228	1	3.797	0
2050	2075	2.981	1	7.642	1	3.119	1	8.149	1	3.248	1	8.755	0
2075	2100	2.889	−1	1.172	−1	3.002	−1	1.317	0	3.181	0	1.480	−1
2100	2125	8.862	−1	2.368	−1	1.709	0	3.612	0	2.870	−1	5.074	2
2125	2150	3.189	−1	1.244	−1	6.608	−1	2.025	−1	1.191	0	3.001	2
2150	2175	1.886	−2	4.146	−2	6.865	−2	8.317	0	1.893	−1	1.415	2
2175	2200	1.168	−2	2.821	−1	7.551	−2	7.959	−1	3.256	−1	1.772	3
2200	2225	9.504	−1	2.866	−1	4.393	0	6.472	0	1.417	2	1.213	3
2225	2250	2.217	2	3.000	2	4.106	2	4.487	1	6.610	3	6.270	2
2250	2275	4.566	3	1.134	3	4.829	3	1.341	2	5.097	4	1.653	3
2275	2300	7.964	3	2.011	3	9.287	3	2.675	2	1.283	5	3.628	4
2300	2325	1.055	5	5.880	5	1.482	5	7.552	2	1.930	5	9.312	5
2325	2350	5.587	5	1.206	5	5.412	5	1.267	3	5.281	4	1.348	5
2350	2375	6.819	4	1.182	4	6.796	4	1.225	3	6.793	−1	1.278	4
2375	2400	1.256	−2	8.873	1	2.320	1	1.249	2	3.640	−1	1.617	−1
2400	2425	7.065	−2	3.404	−1	2.524	−1	6.465	−1	6.360	−1	1.040	−1
2425	2450	8.522	−2	4.236	−1	2.955	−1	7.922	−1	7.297	−1	1.261	−1
2450	2475	1.346	−2	2.405	−1	5.986	−2	4.569	−1	1.927	−2	7.630	−1
2475	2500	4.765	−2	4.248	−1	4.907	−2	4.598	−1	5.684	−1	5.231	−1
3100	3200	1.324	−1	9.836	−1	1.593	−1	1.167	0	1.955	−1	1.349	0
3200	3300	7.731	−2	4.900	−1	1.116	−1	6.244	−1	1.531	−1	7.563	−1
3300	3400	1.232	0	2.952	0	1.440	0	3.374	0	1.666	0	3.792	0
3400	3500	5.159	0	7.639	0	8.430	0	1.124	1	1.284	1	1.601	1
3500	3600	4.299	3	1.914	2	5.007	3	2.321	2	5.757	3	2.779	2

185

Carbon dioxide (continued)

Temperature Wavenumber interval	220 K S	220 K R	260 K S	260 K R	300 K S	300 K R
3600–3700	1.543 (4)	3.245 (2)	1.577 (4)	3.505 (2)	1.618 (4)	3.831 (2)
3700–3800	1.649 (4)	2.722 (2)	1.634 (4)	3.012 (2)	1.646 (4)	3.364 (2)
3800–3900	1.180 (−1)	9.535 (−1)	3.642 (−1)	1.510 (0)	8.943 (−1)	2.260 (0)
3900–4000	1.464 (−2)	2.601 (−1)	1.409 (−2)	2.532 (−1)	1.372 (−2)	2.488 (−1)
4000–4100	1.251 (−2)	2.021 (−1)	2.380 (−2)	2.672 (−1)	4.219 (−2)	3.384 (−1)
4500–4600	9.976 (−3)	2.559 (−1)	1.357 (−2)	3.022 (−1)	1.789 (−2)	3.432 (−1)
4600–4700	2.185 (−1)	1.916 (0)	2.290 (−1)	2.026 (0)	2.433 (−1)	2.132 (0)
4700–4800	2.040 (0)	6.475 (0)	3.437 (0)	8.577 (0)	5.522 (0)	1.105 (1)
4800–4900	1.197 (2)	3.112 (1)	1.220 (2)	3.314 (1)	1.249 (2)	3.525 (1)
4900–5000	4.829 (2)	5.759 (1)	4.880 (2)	6.340 (1)	4.977 (2)	7.006 (1)
5000–5100	8.778 (1)	2.012 (1)	9.629 (1)	2.231 (1)	1.058 (2)	2.460 (1)
5100–5200	8.346 (−1)	1.804 (−1)	8.699 (−1)	2.063 (−1)	9.187 (−1)	2.332 (−1)
5200–5300	8.518 (−2)	8.474 (−1)	1.227 (−1)	1.122 (0)	1.734 (−1)	1.390 (−1)
5300–5400	4.951 (−1)	1.597 (0)	4.943 (−1)	1.639 (−1)	4.969 (−1)	1.676 (0)

Ozone

Temperature Wavenumber interval	220 K S	220 K R	260 K S	260 K R	300 K S	300 K R
0–25	5.791 (2)	1.352 (2)	4.678 (2)	1.187 (2)	3.861 (2)	1.054 (2)
25–50	2.012 (3)	2.923 (2)	1.746 (3)	2.759 (2)	1.535 (3)	2.603 (2)
50–75	2.056 (3)	2.556 (2)	2.032 (3)	2.563 (2)	1.950 (3)	2.518 (2)
75–100	8.957 (2)	1.572 (2)	1.125 (3)	1.802 (2)	1.292 (3)	1.955 (2)
100–125	1.834 (2)	6.041 (1)	3.153 (3)	8.070 (2)	4.554 (2)	9.783 (2)
125–150	1.666 (2)	1.545 (1)	4.401 (1)	2.518 (1)	8.686 (1)	3.519 (1)
900–925	2.622 (−4)	5.784 (−3)	2.045 (−3)	1.549 (−2)	8.913 (−3)	3.121 (−2)
925–950	5.162 (−2)	1.838 (−1)	1.839 (−1)	4.040 (−1)	4.844 (−1)	7.212 (−1)
950–975	2.220 (1)	3.345 (0)	6.569 (1)	5.873 (0)	1.621 (2)	9.121 (0)
975–1000	1.623 (3)	3.521 (2)	3.107 (3)	4.834 (2)	5.054 (3)	6.124 (2)
1000–1025	3.446 (4)	1.397 (3)	3.862 (4)	1.541 (3)	4.157 (4)	1.654 (3)
1025–1050	7.260 (4)	2.207 (3)	6.542 (4)	2.218 (3)	6.013 (4)	2.246 (3)
1050–1075	6.183 (4)	1.652 (3)	6.330 (4)	1.726 (3)	6.398 (4)	1.774 (3)
1075–1100	9.484 (3)	2.204 (3)	1.027 (4)	2.233 (3)	1.051 (4)	2.244 (3)
1100–1125	1.074 (3)	1.755 (2)	1.002 (3)	1.711 (3)	9.843 (2)	1.666 (2)
1125–1150	9.191 (2)	1.661 (2)	1.010 (3)	1.710 (3)	1.089 (3)	1.735 (2)
1150–1175	3.287 (2)	8.363 (1)	3.654 (2)	8.689 (2)	3.901 (2)	8.827 (1)

Bibliography

The following books and review articles are suggested for further reading. They are approximately in the order of the chapters to which they are particularly relevant (and which are indicated by bracketed numbers at the end of the reference) except that those general works which cover most of the field are listed first.

Matveev, L.T. (1967) *Fundamentals of general meteorology: physics of the atmosphere*. Israel Programme for Scientific Translations, Jerusalem.

Hess, S.C. (1959) *Introduction to theoretical meteorology*. Holt, New York.

Hunten, D.M. (1972) Composition and structure of planetary atmospheres. *Space Sci. Rev.* **12**, 539–99. (1)

Goody, R.M. (1964) *Atmospheric radiation*. O.U.P. (2, 4)

Kondratye'v, K. (1969) *Radiation in the atmosphere*. Academic Press. (2, 4)

Brunt, D. (1939) *Physical and dynamical meteorology*. C.U.P. (3)

Iribarne, J.V. & Godson, W.L. (1973) *Atmospheric thermodynamics.*. Reidel, Dordrecht. (3)

Newell, R.E., Vincent, D.G., Dopplick, T.G., Ferruzza, D. & Kidson, J.W. (1970). The energy balance of the global atmosphere. In *The global circulation of the atmosphere*, ed. G. Corby. R. Met. Soc. London. (3, 10)

Dutton, J.R. & Johnson, D.R. (1967) The theory of available potential energy. *Adv. in Geophys.* **12**, 333–436. (3, 10)

Coulson, K.L. (1975) *Solar and terrestrial radiation*. Academic Press. (4)

Akasofu, S. & Chapman, S. (1972) *Solar terrestrial physics*. O.U.P. (5)

Murgatroyd, R. (1970) The physics and dynamics of the stratosphere and mesosphere. *Rep. Prog. Phys.* **33**, 817–80. (5)

Dutsch, H.U. (1973) Recent developments in the photochemistry of atmospheric ozone. *Pure and Appl. Geophys.* **106–8**, 1362–84. (5)

Murata, H. (1974) Wave motions in the atmosphere and related ionospheric phenomena. *Space Sci. Rev.* **16**, 461–525. (5, 8)

COSPAR (1972) *COSPAR International Reference Atmosphere (CIRA)*. Akademie-Verlag, Berlin. (5)

Mason, B.J. (1974) *The physics of clouds*. O.U.P. (6)

Mason, B.J. (1975) *Clouds, rain and rainmaking*, 2nd edn. C.U.P. (6)

Rogers, R.R. (1976) *A short course in cloud physics*. Pergamon. (6)

van de Hulst, H.C. (1957) *Light scattering by small particles*. Wiley. (6)

Feigel'son, E.M. (1966) *Light and heat radiation in stratus clouds*. Israel Programme for Scientific Translations, Jerusalem. (6)

Bibliography

Hansen, J.E. & Travis, L.D. (1974) Light scattering in planetary atmospheres. *Space Sci. Rev.* **16**, 527–610. (6)

Pedlosky, J. (1971) Geophysical Fluid Dynamics. *Lectures in Applied Maths.* **13**, 1–60. (7, 8, 10)

Holton, J.R. (1972) *An introduction to dynamic meteorology.* Academic Press. (7, 8, 9, 10, 11)

Scorer, R.S. (1958) *Natural aerodynamics.* Pergamon. (7, 8, 9)

Charney, J. (1973) Planetary fluid dynamics. In *Dynamic meteorology*, ed. P. Morel. Reidel, Dordrecht. (7, 8, 10)

Chapman, S. & Lindzen, R.S. (1970) *Atmospheric tides.* Reidel, Dordrecht, Holland. (8)

Eckart, C. (1960) *Hydrodynamics of oceans and atmospheres.* Pergamon. (8)

Hines, C.O. & colleagues (1974) *The upper atmosphere in motion.* Amer. Geophys. Union, Geophys. Monograph no. 18. (8)

Lumley, J.L. & Panofsky, H.A. (1964) *The spectrum of atmospheric turbulence.* Interscience. (9)

Sheppard, P.A. (1970) The atmospheric boundary layer in relation to large scale dynamics. In *The global circulation of the atmosphere*, ed. G. Corby. R. Met. Soc. London. (9)

Priestley, C.H.B. (1959) *Turbulent transfer in the lower atmosphere.* University of Chicago Press. (9)

Monteith, J. (1973) *Principles of environmental physics.* Edward Arnold. (9)

Lorenz, E.N. (1967) *The nature and theory of the general circulation of the atmosphere.* World Meteorological Organization, Geneva. (10)

Palmén, E. & Newton, C.W. (1969) *Atmospheric circulation systems.* Academic Press. (10)

Starr, V.P. (1968) *Physics of negative viscosity phenomena.* McGraw-Hill. (10)

Hide, R. & Mason, P.J. (1975) Sloping convection in a rotating fluid. *Adv. in Phys.* **24**, 47–100. (10)

Richardson, L.F. (1922) *Weather prediction by numerical processes.* C.U.P. (reprinted by Dover Publications, 1966). (11)

Marchuk, G.T. (1974) *Numerical methods in weather prediction* (translated from Russian). Academic Press. (11)

GARP Publication Series No. 14 (1974) *Modelling for the first GARP global experiment* (can be obtained from World Meteorological Organization, Geneva). (11)

Thompson, P.D. (1961) *Numerical weather analysis and prediction.* MacMillan. (11)

Smagorinsky, J. (1970) Numerical simulation of the global atmosphere. In *The global circulation of the atmosphere*, ed. G. Corby. R. Met. Soc. London. (11)

Haltiner, G. (1971) *Numerical weather prediction.* Wiley. (11)

Schneider, S.H. & Dickinson, R.E. (1974) Climate modelling. *Rev. of Geophys. & Space Phys.* **12**, 447–93. (11, 13)

Houghton, J.T. & Taylor, F.W. (1973) Remote sounding from artificial satellites and space probes of the atmospheres of the earth and the planets. *Rep. Prog. Phys.* **36**, 827–919. (12)

Bibliography

Lamb, H.H. (1972) *Climate present, past and future*. Methuen. (13)
American Meteorological Society (1968) *Causes of climatic change*, ed. J.M.
   Mitchell, Jr. Meteorological Monographs, 8, no. 30. (13)
GARP Publication Series No. 16 (1975) *The physical basis of climate and
   climate modelling* (can be obtained from World Meteorological Organ-
   ization, Geneva). (13)

References to works cited in the text

Air Force Cambridge Research Laboratories (1965). *Handbook of Geophysics and Space Environments.*

Barnett, J. J. (1974) *Q. J. R. Met. Soc.* **100**, 505–30.

Batchelor, G. (1967) *An introduction to fluid dynamics.* C.U.P.

Chapman, S. (1930) *Mem. R. Met. Soc.* **3**, 103.

Charney, J. G. (1973) In *Dynamical meteorology*, ed. P. Morel. Reidel, Dordrecht, Holland.

Charney, J. G. & Drazin, P. G. (1961) *J. Geophys. Res.* **66**, 83–109.

Charney, J., Fjörtoft, R. & von Neumann, J. (1950) *Tellus* **2**, 237–54.

COSPAR (1972) *COSPAR International Reference Atmosphere (CIRA).* Akademie-Verlag, Berlin.

Crane, A. J. (1977) Unpublished D. Phil. Thesis, University of Oxford.

Crutzen, P. J. (1971) *J. Geophys. Res.* **76**, 7311–27.

Defant, A. (1961) *Physical Oceanography*, **1**, 489–92. Pergamon Press.

Ditchburn, R. W. & Young, P. A. (1962) *J. Atmos. and Terres. Phys.* **24**, 127–39.

Dobson, G. M. B. (1914) *Q. J. R. Met. Soc.* **40**, 123–35.

Dütsch, H. U. (1968) *Q. J. R. Met. Soc.* **94**, 483–97.

Dütsch, H. U. (1971) *Adv. in Geophys.* **15**, 219–322.

Dutton, J. A. & Johnson, D. R. (1967) *Adv. in Geophys.* **12**, 333–436.

Eady, E. J. (1949) *Tellus* **1**, 33–52.

Fjörtoft, R. (1953) *Tellus* **5**, 225.

Gerbier, N. & Berenger, M. (1961) *Q. J. R. Met. Soc.* **87**, 13–23.

Goldman, A. (1968) *J. Quant. Spectrosc. and Radiat. Transfer* **8**, 829–31.

Goody, R. M. & Walker, J. C. G. (1972) *Atmospheres.* Prentice-Hall.

Hanel, R. A., Schlachman, B., Rodgers, D. & Vanous, D. (1971) *Appl. Opt.* **10**, 1376–82.

Harwood, R. S. (1975) *Q. J. R. Met. Soc.* **101**, 75–94.

Harwood, R. S. & Pyle, J. A. (1975) *Q. J. R. Met. Soc.* **101**, 723–48.

Hide, R. (1974) *Proc. R. Soc. London A* **336**, 63–84.

Hide, R. & Mason, P. J. (1975) *Adv. in Phys.* **24**, 47–100.

Holloway, J. L. & Manabe, S. (1971) *Mon. Weather Rev.* **99**, 335–70.

Houghton, J. T. (1963) *Q. J. R. Met. Soc.* **89**, 319–31.

Houghton, J. T. (1972) *Bull. Amer. Met. Soc.* **53**, 27–8.

Houghton, J. T. & Smith, S. D. (1966) *Infra-Red Physics.* O.U.P.

Houghton, J. T. & Smith, S. D. (1970) *Proc. R. Soc. London A* **320**, 23–33.

Inn, E. C. Y. & Tanaka, Y. (1953) *J. Opt. Soc. Amer.* **43**, 870.

Ingersoll, A. P. (1969) *J. Atmos. Sci.* **26**, 1191–8.

References

Jeans, J. H. (1940) *Kinetic theory of gases.* C.U.P.

Jeffreys, H. (1925) *Q. J. R. Met. Soc.* **51**, 347–56.

Julian, P. R., Washington, W. M., Hembree, L. & Ridley, C. (1970) *J. Atmos. Sci.* **27**, 376–87.

Kennedy, J. S. (1964) *Energy generation through radiation processes in the lower stratosphere.* M.I.T. Report, no. 11, Boston.

Krueger, A. J., Heath, D. E. & Mateer, C. L. (1973) *Pure and Appl. Geophys.* **106**–8, 1254–63.

Labitzke, K. and collaborators (1972) *Climatology of the stratosphere – the Northern Hemisphere Part I.* Institut für Meteorologie der Freien Universität Berlin. Meteorologishe Abhandlungen Band 100/Heft 4.

Leovy, C. B. (1969) *Appl. Opt.* **8**, 1279–86.

Lindzen, R. S. (1971) In *Mesospheric models and related experiments*, ed. G. Fiocco. Reidel, Dordrecht, Holland.

Lorenz, E. (1955) *Tellus* **7**, 157–69.

Lorenz, E. (1967) *The nature and theory of the general circulation of the atmosphere.* World Meteorological Organization, Geneva.

McClatchey, R. A., Benedict, W. S., Clough, S. A., Burch, D. E., Calfee, R. F., Fox, K., Rothman, L. S. & Garing, J. S. (1973) *AFCRL atmospheric absorption line parameters compilation.* Air Force Cambridge Research Laboratories Environmental Research Papers no. 434.

McClatchey, R. A. & Selby, J. E. A. (1972) *Atmospheric transmittance 7–30 μm.* Air Force Cambridge Research Laboratories Environmental Research Papers no. 419.

Malkmus, W. (1967) *J. Opt. Soc. Amer.* **57**, 323–9.

Manabe, S. (1969) *Mon. Weather Rev.* **97**, 739–805.

Manabe, S. & Wetherald, R. T. (1967) *J. Atmos. Sci.* **24**, 241–59.

Matveev, L. T. (1967) *Fundamentals of general meteorology: physics of the atmosphere.* Israel Programme for Scientific Translations, Jerusalem.

Milne, E. (1930) Reprinted in *Selected papers on the transfer of radiation*, ed. D. H. Menzel. Dover 1966.

Mintz, Y. (1961) In *The atmospheres of Mars & Venus.* Publication 944, Ad Hoc Panel on Planetary Atmospheres of Space Science Board, National Academy of Sciences, Washington D.C.

Monin, A. S. (1973) In *Dynamical meteorology*, ed. P. Morel. Reidel, Dordrecht, Holland.

National Academy of Sciences (1975) *Understanding climatic change: a programme for action.* Washington D.C.

Newell, R. E., Kidson, J. W., Vincent, D. G. & Boer, G. J. (1972) *The general circulation of the tropical atmosphere*, vol. 1. M.I.T. Press, Boston.

Newton, C. W. (1970) In *The global circulation of the atmosphere*, ed. G. Corby. R. Met. Soc. London.

Oort, A. H. & Peixoto, J. P. (1974) *J. Geophys. Res.* **18**, 2705–19.

Oort, A. H. & Rasmusson, E. M. (1971) *Atmospheric circulation statistics.* Professional Paper no. 5, National Oceanographic & Atmospheric Administration, Washington, D.C.

Palmén, E. & Newton, C. W. (1969) *Atmospheric circulation systems.* Academic Press.

References

Penndorf, R. (1957) *J. Opt. Soc. Amer.* **47**, 176.

Pivovonsky, M. & Nagel, M. R. (1961) *Table of black-body functions.* MacMillan.

Rasool, S. I. & DeBergh, G. (1970) *Nature, Lond.* **226**, 1037–9.

Rodgers, C. D. (1967) *Q. J. R. Met. Soc.* **93**, 43–54.

Rutherford, D. E. (1959) *Fluid mechanics.* Oliver & Boyd.

Shackleton, N. J. & Opdyke, N. D. (1973) *Quaternary Res.* **3**, 39–55.

Smith, W. L. (1976) *Proceedings of study conference on four-dimensional assimilation.* World Meteorological Organization, Geneva.

Starr, V. P. (1968) *Physics of negative viscosity phenomena.* McGraw Hill.

Thekaekara, M. P. (1973) *Solar Energy,* **14**, 109–27, Pergamon.

Thorne, A. P. (1974) *Spectrophysics.* Chapman & Hall.

Vonder Haar, T. & Suomi, V. (1971) *J. Atmos. Sci.* **28**, 305–14.

Vigroux, E. (1953) *Annales de Phys.* **8**, 709.

Williams, A. P. (1971) Unpublished D.Phil. Thesis, University of Oxford.

Answers to problems and hints to their solution

1.1   Earth: 276 K,
             Venus: 326 K,
             Mars: 224 K,
             Jupiter: 121 K.

1.2 $10.8 \, \text{W m}^{-2}$.

1.3
$$p(z) = p_0 \left(1 - \frac{\alpha z}{T_0}\right)^{Mg/\alpha R}$$

1.4   (1) 19.5 km; (2) 14.2 km (use the result of q 1.3).

1.5   3.07%; $\sim 20\%$.

1.6   Mass of atmosphere $= 5.27 \times 10^{18} \, \text{kg}$.
             Thermal capacity of atmosphere $= 5.30 \times 10^{21} \, \text{J K}^{-1}$.
             Thermal capacity of the oceans $= 5.69 \times 10^{24} \, \text{J K}^{-1}$.

1.7   Using the value for c_p for CO_2 of $834 \, \text{J K}^{-1} \text{kg}^{-1}$ ($\approx \frac{9}{2}R$) and for H_2
             of $14{,}450 \, \text{J K}^{-1} \text{kg}^{-1}$ ($\approx \frac{7}{2}R$),
             Mars: 4.5
             Venus: 10.6
             Jupiter: $1.8 \, \text{K km}^{-1}$.

1.8 $9.64 \, \text{K km}^{-1}$ for damp air compared with $9.76 \, \text{K km}^{-1}$ for dry air.

1.9   Solve equation iteratively or graphically: radius of balloon 13.3 m.

1.10   Period of oscillation $= 9.8$ minutes.

2.2 $\psi_0^* = 1.814$; $\phi = 183 \, \text{W m}^{-2}$; Temperature discontinuity 20.7 K.

2.4 $\tau = 0.14$ days.

2.5   Consider rotation of planet around the sun, as well as about its own
             axis; 1 Venus solar day $= 116$ earth days. $\tau \approx 0.44$ Venus days.

2.6 $\chi_0^* \approx 224$.

3.1 $T^* = T (1 + m/\epsilon)/(1 + m)$.
             The required saturation mixing ratios may conveniently be read
             from the tephigram chart:

                      at 273 K, $T^* - T = 0.64$ K,

                      at 290 K, $T^* - T = 2.14$ K,

             and   at 300 K, $T^* - T = 4.11$ K.

3.5 (1) 300 mb.
     (2) Ascent is everywhere stable for dry air. It is stable for saturated air (if the mixing ratio is increased so that the air is saturated) above 780 mb.
     (3) At 1000 mb, 8.3 g kg^{-1}; at 500 mb, 0.75 g kg^{-1}.
     (4) 2 K.
     (5) 950 mb.
     (6) 650 mb.

3.7 Work done on 1st stage of ascent $=$ 7 J
     Energy released in 2nd stage $\quad = 65$ J.

3.8 The temperature of the final mixture is $18\,^{\circ}$C, and hence, for condensation to occur, mixing ratio of the mixture $= 13.1$ g kg^{-1}. Relative humidity $= 93.7\%$.

3.9 (2) The approximation is that the process is replaced by one which is isentropic.

$$\bar{\theta} \; = \; 28.6\,^{\circ}\text{C}$$

$$\bar{m} \; = \; 18.5 \text{ g kg}^{-1}.$$

Condensation level $\hat{=}$ 940 mb.

3.10 Energy released $= 2800$ J kg^{-1}:
     1.8 W m^{-2}:
     Average solar input 245 W m^{-2}.

3.11 4×10^6 J m^{-2}.

4.8 (1) 5.68 mb, (2) 21.3 mb, (3) 1.33 mb.

4.9 $h_m = F_s e^{-1/3} g / 2 c_p p_m$, 0.88 K hr^{-1}.

4.10 If $B_\nu(T)$ is proportional to solar flux, i.e. ignoring all dynamics, changes of ozone concentration etc., temperature difference is ~ 7 K.

4.11 Mass mixing ratio of $H_2O = 0.622 e/p$ (3.11).
     (1) 0.924, (2) 0.92.
     For last part use (4.20). Multiply pathlength by 1.66 in transmission term. Assume spectral interval ~ 500 cm^{-1} wide. Answer 1.0 K day^{-1}.

4.13 To perform the integral over s write it as

$$\underset{\epsilon \to 0}{\text{Lt}} \; N_0 \int \int_\epsilon^\infty \frac{1}{s} \exp\left(-\frac{s}{\sigma}\right) ds [1 - \exp(-sf\rho l)]\, dv$$

$$= N_0 \int \left[\int_\epsilon^\infty \frac{1}{s} \exp\left(-\frac{s}{\sigma}\right) ds - \int_{\epsilon'}^\infty \frac{1}{s'} \exp\left(-\frac{s'}{\sigma}\right) ds' \right] dv$$

where $\epsilon' = \epsilon(1 + \sigma f \rho l)$

5.1 Assume mixing ratio of He at 120 km same as at the surface (Appendix 3) and that oxygen completely dissociated at 120 km. For temperature of 800 K, 770 km; for 1000 K, 940 km; for 1400 K, 1260 km.

Answers and hints

5.2   If T is orbital period, $\Delta T/T = -1.3 \times 10^{-7}$ per orbit;

5.3 $8.3 \times 10^{-10} \text{ m s}^{-1}$.

5.5 $1.36 \times 10^{11} \text{ m}^{-3}$; $2.2 \times 10^{11} \text{ m}^{-2} \text{ s}^{-1}$.

5.7   The resulting temperature profile rises sharply above 120 km, then flattens out so that above 400 km the temperature is substantially constant at about 1250 K.

5.8   Total potential energy, 130 J m^{-2}.
       Power absorbed (from information in problem 5.7), 80 J m^{-2}. Thus we might expect $\Delta T/T$, ~ 0.6 compared with ~ 0.3 from fig. 5.4.

5.9 $(J_3 + k_2 n_2 n_M)^{-1}$; $\dfrac{J_3 + k_2 n_2 n_M}{4n_2(J_2 J_3 k_2 k_3 n_M)^{1/2}}$.

       First time constant: 0.76 s at 40 km; 0.05 s at 30 km; and 2×10^{-3} s at 20 km.
       Second time constant: 1.6×10^5 s at 40 km; 1.3×10^6 s at 30 km: 1.3×10^7 s at 20 km.

5.10 0.4 K day^{-1}.

5.12   0.023.

5.13 $g_2/g_1 = 2$ for $15 \mu m$ CO$_2$ band; cooling rate $\hat{=} 5 \text{ K day}^{-1}$.

5.14 90 K day^{-1}.

5.15   Transmission at line centre $\hat{=} 0.66$.

5.16 7.5×10^9: 2.2×10^{-5}.

6.3   (1) 26.5 min; (2) 7.34 hr.

6.4   91.9 days; 22 hours; 13.2 min.

6.5   22.9 min; $-9.6\,°$C.

6.6 $A = 2/3, \tau = 1/3$.

6.9 $A = 0.6635, \tau = 0.3353, (1 - A - \tau) = 5.97 \times 10^{-3}$.

6.10   19.56

6.11 $\beta = 0.6138, \chi_0^* = 34.4$.

6.12 $\beta = 0.0266$, with $\chi_0^* = 4, \tau = 0.067$.

6.13   0.9972.

6.14   At $0.3 \mu m$, 0.637, at $0.6 \mu m$, 0.061, at $1 \mu m$, 0.008.

7.1 $\Omega^2 R = 3.457 \times 10^{-3} g$: 5.95 arc minutes.

7.3   Magnitudes of terms (cm s^{-2}):

       $$\frac{du}{dt}, \frac{dv}{dt} \hat{=} 10^{-2}, \frac{uv \tan \phi}{a}, \frac{u^2 \tan \phi}{a} \hat{=} 10^{-3},$$

       $$2\Omega u \sin \phi, 2\Omega v \sin \phi \hat{=} 10^{-1}, 2\Omega w \cos \phi \hat{=} 10^{-4},$$

       $$uw/a, vw/a \hat{=} 10^{-6}.$$

7.4   3.9 mb/100 km.

7.5 7.29 m s^{-1}.

7.6   1 day at $30°$ latitude.

195

Answers and hints

7.7   0.194. On Mars, 0.20; on Venus, 47.1.

7.8 $\sim 10\%$.

7.11 $g = g_0 a^2 (a + z)^{-2}$.

7.12 19 m s^{-1}.

7.13   Consider pressures on either side of the frontal surface, p_1, p_2, and write

$$\delta p_1 = \frac{\partial p_1}{\partial x} \delta x + \frac{\partial p_1}{\partial z} \delta z$$

and similarly for δp_2. At front, $p_1 = p_2$, $\delta p_1 = \delta p_2$ and $\tan \alpha = \delta z / \delta x$.

$$\alpha = 0.41° \text{ at } 30° \text{ latitude}.$$

7.15   Temperature will fall.

7.16 $V_{200\,\text{mb}} = 17.9 \text{ m s}^{-1}$ at $45°$ latitude.

7.17   Average temperature gradient, 1 K per 100 km: wind speed, 33.5 m s^{-1}.

7.25   Transformations of the form derived in q 7.24 are applied to the continuity equation.

7.26   For scalar ϕ and vector $\mathbf{F}$, curl $\phi \mathbf{F} = \phi$ curl $\mathbf{F} + \nabla \phi \wedge \mathbf{F}$. Also curl $\nabla \phi = 0$, and g is the gradient of a scalar potential.

8.4 $1.88 \times 10^{-2} \text{ s}^{-1}$.

8.5 $\lambda_h \simeq 10$ km, $\lambda_v \simeq 1$ km, horizontal phase speed $\simeq 16 \text{ m s}^{-1}$.

8.6 $\omega_a = \omega_B$ if there is 2.2 K km^{-1} temperature inversion.

8.8   The kinetic energy density is comprised of terms of form $\frac{1}{2}\rho u^2$: 0.14 m s^{-1} if $\bar{T} = 260$ K.

8.9   (1) 3.34 km; (2) 4.78 km.

8.10 $(v_g)_h = \omega_B \dfrac{m^2}{(m^2 + k^2)^{3/2}}$, $(v_g)_v = - \dfrac{\omega_B km}{(m^2 + k^2)^{3/2}}$.

8.12 2.67 m s^{-1}.

8.13 2.48 m s^{-1}.

8.14 $v_g = \bar{u} + \dfrac{\beta(k^2 - l^2)}{(k^2 + l^2)^2}$.

8.15 $0.02; 0.01$.

8.16 ζ may be approximated to the vorticity of geostrophic wind to an approximation of order Ro.

8.18 1.5×10^{-5}: 3×10^{-5}.

8.19   The method is as for q 8.10 or 8.14.

8.20   The method is as for q 8.2 or 8.11.

9.3   The effect of buoyancy must be considered. We expect $K_E \simeq K; K_Q$ will be different.

9.5   If $\zeta_g = 10^{-5} \text{s}^{-1}$, $H = 10$ km, $K = 10 \text{ m}^2 \text{s}^{-1}$, $w_d \simeq 7 \text{ mm s}^{-1}$. Spin down time $\simeq 4$ days.

Decay time under damping from eddy viscosity but ignoring vertical
motion and secondary circulation $\triangleq H^2/K$, ~ 100 days.

10.1   (a) 2.94; (c) 0.337; (e) 0.032.

10.3 $\partial u^{(0)}/\partial z \triangleq 2 \text{ m s}^{-1}\text{km}^{-1}$ (using, for example, the result of q 7.16).
Max. values of $u^{(1)}$, $v^{(1)}$, $\sim 0.2 \text{ m s}^{-1}$.

10.4
$$w = \frac{1}{a \cos \phi} \frac{\partial}{\partial \phi}\left[\frac{K}{f} \cos \phi \left(\frac{\partial u}{\partial z}\right)^0 \right]$$

0.02 mm s^{-1}.

10.8   For $c_i \geqslant 0$ r.h.s of (10.33) must be $\geqslant 0$ and using substitution provided
$$\frac{1 + \tanh^2 (H\alpha/2)}{\tanh (H\alpha/2)} \geqslant \frac{1 + H^2\alpha^2/4}{H\alpha/2}$$

10.10   With $\lambda_m = 4000 \text{ km}$, $H = 10 \text{ km}$, $\partial \bar{u}/\partial z = 2 \text{ m s}^{-1}\text{km}^{-1}$,
$(kc_i)^{-1} \triangleq 2.10^5 \text{ s}$.

10.13 62.2 m s^{-1}.

10.14 2×10^{-6}: 4 m s^{-1}.

12.1   (a) 0.3%; (b) 3%; (c) 3%;

12.2   Use method of q 4.11: $\sim 1.8 \text{ K}$.

12.5   The following theorem concerning differentiation under the integral
sign will be found useful:

if $F(\beta, \alpha) = \int_{\alpha}^{\beta} f(x, a)\, dx$

$$\frac{dF}{da} = f(\beta, \alpha) \frac{d\beta}{da} - f(\alpha, a) \frac{d\alpha}{da} + \int_{\alpha}^{\beta} \frac{\partial f(x, a)}{\partial a}\, dx$$

12.6   (a) 5.8%; (b) 1.7%; (c) 0.4%.

12.8   0.25 K.

12.10   210 km.

12.11   Scattering coeff. $\sim 4.10^{-10} \text{ cm}^{-1}$; absorption coeff. 10^{-7}cm^{-1}.

12.12   (a) 9.5 km; (b) 19.9 km.

12.13   Condition for ray to emerge is that horizontal ray distant r from
centre of planet shall have radius of curvature $> r$. From information
in problem 12.12 this condition is found to be

$$-\frac{dn}{dr} < \frac{n}{r}$$

For Venus atmosphere, assuming temperature of $\sim 410 \text{ K}$, highest
pressure 3000 mb.

Index

absorption
  bands, 33–6, 179–88
  by carbon dioxide, 140–1, 186–8
  coefficient, 8–10, 33
  cross-section, 57
  spectra, 8–9, 34
  spectrum of oxygen, 180–1
  spectrum of ozone, 38, 180–1, 188
  by water vapour, 34, 140–1, 182–5
absorptivity, of clouds, 70, 73
acoustic cut-off frequency, 90
acoustic waves, 90, 137
adiabatic
  lapse rate, 3–4
  super-adiabatic condition, 5
  wet, 17–21
air, thermodynamic data for, 164
airglow, 56, 65
albedo, 2, 41
  of clouds, 70
  for single scattering, 70
angular momentum
  of atmosphere, 133
  transport of, 128–9
  zonal, 26
anticyclone, 78, 132
argon, 47, 166
atmosphere
  composition of, 1, 166
  model atmospheres, 168–76
  plane parallel, 9
  predictability of, 160
  unit of pressure in, 1
  upper, 46–66, 176
available potential energy, 22–7, 115, 130
  eddy, 26, 115, 130
  generation of, 26, 30
  zonal, 26, 115, 130

back-scatter ultraviolet spectrometer, 155
balloons, 7, 147
  constant density, 155
band
  electronic, 8, 180–1
  models, 39–40

pure rotation, 8, 182–8
  strength, 57
  strong, 8
  vibration–rotation, 8, 33, 182–8
baroclinic
  instability, 115, 123–7
  model, 134
  waves, 115
barotropic
  atmosphere, 134
  instability, 122–3
  model, 134
beta-plane approximation, 94, 97, 123
Bjerknes–Jeffreys theorem, 86
black-body radiation, 8, 10
  from sun, 8, 179
black-body function (see also Planck function), 61
Boltzmann factor, 57
Bouguet's law, 9
boundary layer, 104, 105–7
  in numerical models, 143–4
Boussinesq approximation, 96, 97, 117
Bowen ratio, 111
broadening
  collision, 33, 37, 42, 44, 45
  Doppler, 33, 42, 66
Brunt–Vaisala frequency, 7, 88, 90, 99

carbon dioxide
  absorption by, 34, 140–1, 184–6
  and climatic change, 162
  collisional relaxation in, 61, 66
  concentration of, 166
  emission, 37–9, 57–63, 149, 151
  and non LTE, 61
centrifugal acceleration, 76, 82
cirrus clouds, 67
Clausius Clapeyron equation, 19, 72
climate
  of past, 161
  variations, 161
  modelling, 162
climatic change, 1, 160–3
clouds, 67–73
  cirrus, 67
  cumulus, 67

emissivity of, 70, 143
formation of, 67
interaction with radiation 40−1,
    69−72, 142−3
lenticular or wave, 67, 92
in numerical models 142−3
particles, 67−9
satellite pictures of, 147−8
stratus, 67, 70, 72−3
on Venus, 2, 73
coalescence, in clouds, 67−9
collision
    broadening, 33, 37, 42, 44, 45
    excitation of molecules by, 57
    relaxation by, 57, 61, 66
composition
    of atmosphere, 166
    of thermosphere, 47
computational instability, 137
condensation
    and formation of precipitation, 67−9
    level, 18
constant flux layer, 107
continuity, equation of, 81, 86, 89
continuum absorption, 40, 43
convection, 8, 12−13
    in numerical models, 138
    in a rotating fluid, 115
    sloping, 115, 117, 127
convective adjustment, 138
cooling to space approximation, 38, 61,
    62
Coriolis parameter, variation with lati-
    tude, 93, 101−2
Coriolis term, 76, 78, 79, 83
COSPAR reference atmosphere, 51, 52,
    54, 176
critical level, 50, 51, 53
cumulus clouds, 67
Curtis−Godson approximation, 37, 42
Curtis matrix, 63
curve of growth, 36
cyclone, 78, 132
    mid-latitude, 126
    tropical, 79
cyclostrophic motion, 79

Dalton's law, 47
derivative
    partial or local, 74
    total, 74
dew point, 20
diffusive separation, 47
dishpan experiment, 116
dissociation rate, 55
Doppler broadening, 33, 42, 66
drag coefficient, 63, 144, 145

Eady wave, 126, 133

Eckman layer, 108, 118, 119
Eckman pumping, 108−9
Eckman spiral, 106−7
eddy
    available potential energy, 26, 115,
        130
    diffusion coefficients, 143
    kinetic energy, 26, 127
    stresses, 105−6
    transfer coefficients, 112, 143
    viscosity, 105
Einstein coefficients, 58−60
Elsasser band model, 39, 150, 157
emission of radiation, 8−9
    by carbon dioxide, 37−9, 57−63,
        149, 151
    by clouds, 70
    by planet, 2
    stimulated and spontaneous, 57−8
    by water vapour, 141
emissivity
    of water vapour, 141
    of clouds, 70, 143
    of surface, 143, 148
energy
    available potential, 22−7, 115, 130
    budget, 33, 128
    eddy kinetic, 26, 127, 130
    internal, 20, 22
    kinetic, 22, 25
    total potential, 20−3, 24, 25
    transport, 128
    zonal kinetic, 26, 130
entropy, 17−21
    of dry air, 17
equation
    of continuity, 81, 82, 86, 89, 119
    of motion, 77
    of state, 3, 17, 18, 20
    thermal wind, 80, 85, 117, 125
equivalent width, 35−9, 44
    of collision broadened line, 42
exosphere, 50

feedback processes in atmosphere, 162
Fick's law of diffusion, 68
finite differences, 136
flux
    of hydrogen atoms, 53
    radiative, 10, 33
    solar, 2, 39
fog, 28
frame of reference, 74
    inertial, 75
    rotating, 75−6
friction
    effect on geostrophic approx. 78
    in the boundary layer, 108−9
    velocity, 108

front, 84, 85
frost point, 20

general circulation, 26, 115–33
geopotential, 80
    height, 80, 84, 167
geostrophic
    approximation, 77–80,
    wind, 77, 83, 86, 96, 97, 98, 106
    quasi-geostrophic approximation, 98,
        137
    zonal velocity, 121
Global Atmospheric Research Program,
    160
    first global experiment, 146
Goody random model, 39
gradient wind, 79, 83
gravity waves, 7, 88–93, 99, 101, 137.
greenhouse effect, 3, 14
    runaway, 14–15,
    on Venus 3, 14
grey atmosphere approximation, 8, 10

Hadley circulation, 117, 120, 129
Hartley band of ozone, 38
heating rate, radiative, 33, 39, 60–2,
    139
helium, 50, 63, 64, 166
homopause, 47
homosphere, 47
hydrogen, 166
    in the upper atmosphere, 47, 50–3
hydrostatic equation, 3, 28, 89
    for moist air, 18
hydrostatic equilibrium, 22

ice
    in glaciers, 161
    particles in clouds, 68, 73
inertial
    flow, 83
    frame of reference, 75
    instability, 121–2
    oscillations, 99
instability, 121
    baroclinic, 115, 122–4
    barotropic, 122–3
    computational, 137
    inertial, 121–2
    latent, 29
    potential, 29
integral equation of transfer, 31–2
integrated absorptance, 35
internal energy, 20, 22
intertropical convergence zone, 123, 124
intrinsic frequency of gravity waves, 91
inversion, of temperature profile, 5
isentropic surfaces, 22, 23

isobaric
    surface, 23
    system, 82
isobars, 20, 84

jet stream, 85
    polar front, 117
    subtropical, 117
Jupiter
    internal energy, source in, 2, 6
    negative viscosity situation in, 130
    radiative parameters for, 2
    structure of atmosphere, 1

Kelvin waves, 102
kinetic energy, 22, 25
    eddy, 26, 127, 130
    zonal, 26, 130
Kirchhoff's law, 10
Kolmogoroff law, 113

Ladenberg and Reiche function, 42
Lambert's law, 9
laminar flow, 103
lapse rate, 12
    adiabatic, 3–4
latent heat
    of condensation, 18, 139, 165
    of fusion, 165
    of sublimation of ice, 73
latent instability, 29
lenticular clouds, 67
line strength, 35, 37, 180

Malkmus model, 44
Mars
    greenhouse effect for, 14
    negative viscosity situation in, 130
    radiative parameters for, 2
    radiative time constant in, 16
    structure of atmosphere, 1
mass action, law of, 55
mesopause, 46, 56
mesosphere, 37, 46
meteor trails, 147
methane, 51, 166
millibar, unit of pressure, 1
mixing
    in lower atmosphere, 47
    length, 107
    ratio of water vapour, 18–21, 166–7
molecular diffusion, 47, 53
Monin–Oboukhov length, 112

Nimbus
    clouds, 147
    satellite, 150, 151, 152, 153, 156
nitrogen, 47, 50, 166

noise equivalent power (NEP), 158
noise equivalent temperature (NET), 158
nuclei
  for condensation, 67
  hygroscopic, 72
numerical modelling, 134–45
  baroclinic models, 134–6
  barotropic models, 134
  of climate, 163
  data requirements for, 146
  primitive equations for, 136–8
  spectral, 145
  of transfer across the surface, 143–5
  two dimensional, 144

occultation of radio signals by Venus, 159
optical depth, 9, 11, 70
optical path, 9
orography
  cloud formation due to, 67
  in numerical models, 136, 138
oxygen
  absorption by, 46, 53, 180–1
  atomic, 47, 50–1
  isotopic composition of, 161
  microwave band of, 152
  origin of atmospheric, 64
  photochemistry of, 55
ozone, 38–9, 46, 55–6, 65
  absorption by, 38–9, 139, 179–80, 186
  distribution of, 43
  heating rate by, 43
  remote sounding of, 153
  spectral band information for, 186

partial derivative, 74
perturbation method, 87, 89
photochemical processes, 55–6
  in numerical models, 144
photodissociation, 55
physical constants (table of), 164
Planck function (*see also* black-body function), 32, 142, 148, 176–7
planetary waves, 93, 94, 123
polar front jet, 117
polarization of sunlight, 73
potential energy, 20–3
  available, 22–7, 115, 130
  total, 22–5
potential instability, 29
potential temperature, 17, 20, 23, 25, 29, 89
potential vorticity, 96
predictability, atmospheric, 160–3
pressure modulator radiometer, 151, 156
primitive equation models, 136–7

quasi-geostrophic approximation, 98, 137
quasi-horizontal motion, 77

radiance, 149
radiation
  black-body, 8, 10
  budget, 40, 41
  infrared, 2, 8, 147
  solar, 2, 8, 38–41, 139–41, 147, 177–8
  terrestrial, 41, 139, 142
  ultraviolet, 46, 147, 153
radiative
  equilibrium model, 8–16
  time constant, 13, 14, 16
  transfer, 8, 31–45, 63, 139
  transfer in clouds, 70
  transfer in thermosphere, 53
radiosonde, 147
  ascent, 28
random band model, 39, 44
Rayleigh scattering, 70, 73, 158
Rayleigh's criterion, 123
reaction rates, 55, 65
  for three-body collisions, 55, 65
refractive index
  in spherical atmosphere, 159
  of dry air, 164
relative humidity, 18
relaxation, collisional, 57, 61, 66
remote sounding from satellites, 147
  of atmospheric temperature, 147–52
  of atmospheric composition, 152–5
  from atmospheric limb, 152, 154, 15
Reynold's number, 103
Reynold's stresses, 104, 105
Richardson number, 111, 114
  flux Richardson number, 113
rocket sondes, 147
rocket trails, 47, 92, 93
Rossby number, 76, 83, 131
Rossby waves, 93, 95, 96, 102
roughness length, 108

Sändstrom's theorem, 4–6
satellites, 64, 147
  geostationary, 147
  polar orbiting, 147
saturation mixing ratio, 18, 20
saturation vapour pressure, 18, 73
scale height, 3, 51
  in the thermosphere, 47
scattering
  in clouds, 70, 71
  coefficient, 70
  isotropic, 71
Schwarzchild's equation, 10, 31

Index

selective chopper radiometer, 149, 153
sigma co-ordinates, 138, 145
sloping convection, 115–7, 127
solar constant, 2, 177
solar wind, 53
sound waves, 87
source function, 57, 61
spectral
  band information, 180–6
  energy density of turbulence,
    109–11
  lines, 33
  numerical model, 145
spin-down time, 109, 112
stability (*see also* instability)
  for vertical motion of dry air, 4, 5
  of moist air, 19
standard pressure, 17, 164
static stability parameter, 135
steering level for Eady waves, 125
Stefan–Boltzmann constant, 2, 164
Stefan–Boltzmann law, 10
Stokes' law, 72, 73
stratopause, 43, 46
stratosphere, 13, 37, 55
  energy transport in, 129
  temperature structure of, 48–9, 154,
    168–75
  water vapour in, 166
stratospheric warming, 97, 98
stratus clouds, 67, 70, 72, 73
streak photographs, 116
stream function, 94
strong approximation, 35, 37–8, 42, 44
sub grid scale processes, 143
sub-tropical jet, 117
supercooled droplets, 67
supersaturation, 72
symmetric circulation, 116–20
synoptic scale, 77, 83

temperature
  discontinuity at surface, 12, 16
  distribution of atmospheric, 48–9,
    168–75
  entropy plot, 20
  equilibrium 2, 6
  inversion, 5
  of past century, 161
  potential, 17, 20, 23, 29, 89
  virtual, 27
tephigram, 20, 21, 28–30
terrestrial radiation, 41, 139, 142
thermal wind equation, 80, 85, 117, 125
thermodynamic
  breakdown of LTE, 57
  cycle, 5, 6
  diagram, 20
  engine, 5

equation, 135
equilibrium, 9, 10
  local thermodynamic equilibrium
    (LTE), 10, 57, 60, 62
thermodynamics, 17–30
  first law of, 4, 18
thermosphere, 46, 47, 53
thickness, 80
tornado, 83
total derivative, 74
transmission
  of an atmospheric path, 37, 40, 63
  average, 33, 36–9
  of clouds, 70, 73
  fractional, 9
  of slab, 38
transmissivity, of clouds, 70, 73
tropical cyclone, 79
tropopause, 13, 46
troposphere, 13
tropospheric forcing, 129
turbopause, 47, 53
turbulence, 103–14
  spectrum of atmospheric, 109, 110
  three-dimensional, 109
  two-dimensional, 110

Venus
  cloud albedo of, 73
  density measurements for, 159
  cloud cover of, 2
  greenhouse effect on, 3, 14–15
  lower atmosphere of, 6, 120–1
  radiative parameters for, 2
  structure of atmosphere of, 1, 15
virtual temperature, 27
viscosity
  eddy, 105
  kinematic, 103
  molecular, 103, 105
  negative, 130
von Karman's constant, 107, 112
vorticity, 93, 96, 102, 108
  absolute, 93, 94, 122
  equation, 86, 95, 124, 134–6
  potential, 96
water vapour
  absorption by, 34, 140–1, 180–3
  continuum, 40, 43
  dimer, 40
  distribution, 167
  emissivity of, 141
  and greenhouse effect, 14–5
  in numerical models, 138–9
  properties of, 165–6
  spectrum, 34, 177
  in upper atmosphere, 51, 166
  and vertical motion of moist air, 17
wave clouds, 67

wave equation, 88
wavenumber, 32, 97
waves
   acoustic, 90, 137
   atmospheric, 87–102
   baroclinic, 115
   Eady, 126, 133
   external, 90
   gravity, 7, 88–93, 99, 101, 137
   internal, 90
   Kelvin, 102
   lee, 91, 92
   planetary, 93–4, 123
   Rossby, 93–5, 102
weak approximation, 35, 37, 42, 44
weighting function, 149, 150, 157
wet adiabatic, 20, 21

wind
   in boundary layer, 104–7
   geostrophic, 77, 83, 86, 96, 97, 98,
      106
   gradient, 83
   mean zonal, 50
   thermal, 80
   in thermosphere, 92
window region, 41, 148, 149

zenith angle, solar, 39
zonal
   angular momentum, 26
   available potential energy, 26, 115,
      130
   kinetic energy, 26, 130